Surfaces

Gary Attard

Senior Lecturer, Department of Chemistry, University of Wales, Cardiff

Colin Barnes

Lecturer, School of Chemical Sciences, Dublin City University

OXFORD
UNIVERSITY PRESS
Great Clarendon Street, Oxford OX2 6DP
Oxford University Press is a department of the University of Oxford.
It furthers the University's objective of excellence in research, scholarship,
and education by publishing worldwide in

Oxford New York

Auckland Cape Town Dar es Salaam Hong Kong Karachi
Kuala Lumpur Madrid Melbourne Mexico City Nairobi
New Delhi Shanghai Taipei Toronto

With offices in
Argentina Austria Brazil Chile Czech Republic France Greece
Guatemala Hungary Italy Japan Poland Portugal Singapore
South Korea Switzerland Thailand Turkey Ukraine Vietnam
Oxford is a registered trade mark of Oxford University Press
in the UK and in certain other countries

Published in the United States
by Oxford University Press Inc., New York

First published 1998
Reprinted 2001, 2003, 2004, 2006

British Library Cataloguing in Publication Data
Data available

Library of Congress Cataloging in Publication Data
Attard Gary
Surfaces / Gary Attard, Colin Barnes.
(Oxford Chemistry primers: 59)
Includes bibliographical references and index.
1. Surface chemistry. 2. Surfaces (Physics). 3. Surfaces
(Technology) I. Barnes, Colin. II. Title. III. Series
QD506.A88 1998 541.3'3—dc21 97-43823

ISBN 13: 978-0-19-855686-2
ISBN 10: 0-19-855686-1

7 9 10 8 6

Printed in Great Britain on acid-free paper by
Antony Rowe Ltd, Chippenham, Wiltshire

Series Editor's Foreword

Oxford Chemistry Primers are designed to provide clear and concise introductions to a wide range of topics that may be encountered by chemistry students as they progress from the freshman stage through to graduation. The Physical Chemistry series aims to contain books easily recognised as relating to established fundamental core material that all chemists need to know, as well as books reflecting new directions and research trends in the subject, thereby anticipating (and perhaps encouraging) the evolution of modern undergraduate courses.

In this Physical Chemistry Primer Gary Attard and Colin Barnes present a logical, simply written and stimulating account of the fundamental concepts of *Surface Science* and the methods used to investigate the properties of interfaces; a topic of intense research activity—both academic and industrial. This Primer will be of interest to all students of chemistry and their mentors.

Richard G. Compton
Physical and Theoretical Chemistry Laboratory
University of Oxford

Preface

The chemistry and physics of surfaces is an increasingly important subject. This is a consequence of the many technological applications to which the subject lends itself—heterogeneous catalysis, corrosion, printing, dyeing, detergency, and adhesion, to name but to a few. The study of surfaces is also making a significant contribution to our understanding of fundamental areas of science, including phase transitions, electronic structure, and chemical bonding. Because of this, there is a need for chemists to become familiar with the basic concepts and principles governing interfacial phenomena. To this end, the present, self-contained, introductory account of the subject of surface science is addressed. It will first seek to explain the singular behaviour of solid surfaces within a framework based on a number of key macroscopic and measurable parameters. The concept of **adsorption** will be emphasized and its quantification through thermodynamic and spectroscopic methods will be described. Of necessity in a primer of this size, we have been unable to include all of the areas of modern surface science research that we would have liked but is has been our aim to provide final year undergraduates/first year postgraduate students with a solid basis on which to proceed. In particular, the inclusion of a chapter devoted to worked examples is used to illustrate the various types of molecular physicochemical information obtainable from a multi-technique approach to surface analysis and to highlight concepts introduced in earlier chapters.

Finally, may we extend our sincere thanks to our research students (and also John Freeman at Oxford) for their assistance in producing the diagrams and Pat Regan for her speedy and efficient typing of the manuscript.

Cardiff G.A.
Dublin C.B.
1997

Contents

1 Gas adsorption at solid surfaces

Technological developments over the last 80 years or so (from light bulbs to three-way automobile catalytic converters) have provided the impetus for intensive studies of gas–solid interfaces. The central importance of this subject to many areas of pure and applied research (corrosion, electronic device manufacture, heterogeneous catalysis) cannot be underestimated. In this chapter, some key ideas relating to the interaction of gases with solids will be described.

1.1 Some basic definitions

Adsorption is the term used to describe the process whereby a molecule (an *adsorbate*) forms a bond to the surface (the *adsorbent*). It should be distinguished from **absorption**, which refers to molecules entering into the bulk of the substrate. The **fractional coverage** of adsorbate (usually given the symbol, θ) is defined as

$$\theta = \frac{\text{Number of surface sites occupied by adsorbate } (N_s)}{\text{Total number of substrate adsorption sites* } (N)} \tag{1.1}$$

*N is often numerically equivalent to the total number of surface atoms of the substrate.

When $\theta = 1$, the adsorbate ensemble is called a **monolayer**.

Associative adsorption is said to occur when a molecule adsorbs on to the surface from the gas phase without fragmentation. When fragmentation does occur, the adsorption process is termed **dissociative**. The types of bonding interactions formed between gas molecules and solid surfaces, and the rate at which vacant adsorption sites are filled, will be discussed in Sections 1.5 and 1.6. In what follows, a simple, elegant model describing the equilibrium between the gaseous and adsorbed phase will be developed. It is based on ideas proposed by one of the most distinguished pioneers of surface science, Irving Langmuir.

1.2 The Langmuir adsorption isotherm

The number of surface sites occupied by adsorbate molecules at equilibrium at a particular temperature will depend on the gas pressure, P. The dependence of θ on P at constant temperature is called an **adsorption isotherm**. At sufficiently low values of P all adsorption isotherms are **linear** and may be regarded as obeying Henrys' law

$$P = \text{constant} \times \theta \tag{1.2}$$

The **Langmuir isotherm** has been used successfully to interpret equilibrium adsorption behaviour of a number of systems and in determining the **total surface area**, S_A, of solid surfaces, particularly high surface area solids often used as heterogeneous catalysts. As with all models, some often severe approximations have to be made which may not always reflect the reality of the situation:

(i) The solid surface is uniform and contains a number of *equivalent* sites each of which may be occupied by only one molecule of adsorbate.

(ii) A dynamic equilibrium exists between the gas (at pressure, P) and the adsorbed layer at constant temperature.

(iii) Adsorbate molecules from the gas phase are continually colliding with the surface. If they impact a vacant adsorption site, they form a bond with the surface and stick. If they strike a filled site, they are reflected back into the gas phase.

(iv) Once adsorbed, the molecules are localized (that is, the activation barrier hindering migration to an adjacent site is much greater than kT; see Section 1.12) and the enthalpy of adsorption per site remains constant irrespective of coverage.

k = Boltzmann constant
T = Absolute temperature

First, let us assume that the molecules in the gas phase are in dynamic equilibrium with the surface,

$$M_{(g)} + S_{(\text{surface site})} \underset{k_d}{\overset{k_a}{\rightleftharpoons}} M-S \quad \text{(Associative Adsorption)} \tag{1.3}$$

where k_a and k_d are the rate constants for the **adsorption** and **desorption** steps, respectively. If P is the pressure and θ is the fractional monolayer coverage of the surface by adsorbate molecules

$$\text{rate of adsorption} = k_a P(1-\theta) \tag{1.4}$$

where $(1-\theta)$ is the fractional monolayer coverage of sites *not* occupied by adsorbate molecules. Equation 1.4 implies that the rate of adsorption will be fast if k_a and P are large and θ is small. Similarly,

$$\text{rate of desorption} = k_d\theta \tag{1.5}$$

Note that, according to eqn 1.5, the rate of desorption will be independent of the pressure, P, but will depend on θ. None the less, P does determine θ so there is an 'indirect' influence on the rate of desorption. At equilibrium, both the rate of adsorption and desorption are equal

$$k_a P(1-\theta) = k_d\theta \tag{1.6}$$

and upon rearrangement of eqn 1.6

$$\theta = \frac{N_s}{N} = \frac{KP}{1+KP} \qquad \left(K = \frac{k_a}{k_d}\right) \tag{1.7}$$

K is the equilibrium constant corresponding to eqn 1.3

K should be distinguished from the **thermodynamic** equilibrium constant $K^0 = KP^0$, where P^0 = standard pressure.

Equation 1.7 is the **Langmuir adsorption isotherm** for associative adsorption and predicts how the fractional monolayer coverage θ of adsorbate changes with P. An analogous equation may be derived for dissociative adsorption

$$\mathrm{M_{2(g)}} + 2\mathrm{S} \underset{k'_d}{\overset{k'_a}{\rightleftharpoons}} 2(\mathrm{M-S}) \tag{1.8}$$

$$\text{rate of adsorption} = k'_a P(1-\theta)^2 \tag{1.9}$$

[two sites are required for dissociative adsorption hence rate is second order in $(1-\theta)$]

$$\text{and rate of desorption} = k'_d \theta^2 \tag{1.10}$$

[two adsorption sites are occupied hence desorption requires probability of two molecules on surface reacting and rate is second order in θ]

At equilibrium

$$\text{rate of adsorption} = \text{rate of desorption}$$

$$\therefore \quad k'_a P(1-\theta)^2 = k'_d \theta^2$$

$$\therefore \frac{\theta^2}{(1-\theta)^2} = \frac{k'_a P}{k'_d} = K'P \qquad K' = \frac{k'_a}{k'_d} \tag{1.11}$$

(K' is the equilibrium constant for reaction 1.8)

$$\therefore \quad \frac{\theta}{1-\theta} = (K'P)^{\frac{1}{2}}$$

$$\Rightarrow \theta = \frac{(K'P)^{\frac{1}{2}}}{1+(K'P)^{\frac{1}{2}}} \tag{1.12}$$

Two features of eqn 1.7 should be noted

$$\text{as } P \overset{\lim}{\longrightarrow} 0, \left(\frac{KP}{1+KP}\right) = 0 \tag{1.13}$$

and, as expected when $KP << 1$ (low P),

$$\theta = \frac{KP}{1+[\text{small number}]} = KP \tag{1.14}$$

linear dependence of θ on P (Henry's law approximation; eqn 1.2)

In contrast

$$\text{as } P \overset{\lim}{\longrightarrow} \infty, \left(\frac{KP}{1+KP}\right) = 1 = \theta \tag{1.15}$$

Equation 1.15 is the condition when all adsorption sites are filled with adsorbate—the formation of a complete monolayer. The equilibrium constant, K, represents the affinity of a particular molecule for a surface. Large values of K imply that a strong bond is formed between the adsorbate and the substrate, whereas for small K the opposite is true. Since K is an equilibrium constant, it may be used to calculate other thermodynamic data pertaining to the adsorption system. Figure 1.1 summarizes pictorially the essential features of the Langmuir adsorption isotherm.

The coverage θ may be defined not only in terms of relative numbers of molecules (eqn 1.1) but also in terms of relative masses and relative volumes

$$\theta = \frac{N_s}{N} = \frac{m}{m_\infty} = \frac{V}{V_\infty} = \frac{KP}{1+KP} \tag{1.16}$$

m = mass of gas adsorbed; m_∞ = mass of gas adsorbed corresponding to all adsorption sites being occupied in monolayer; V = volume of gas adsorbed at constant P; V_∞ = volume of gas adsorbed at constant P corresponding to all adsorption sites being occupied.

Taking eqn 1.7 and multiplying out

$$NKP = N_s + N_s KP \tag{1.17}$$

$$\left(\frac{P}{N_s}\right) = \frac{1}{NK} + P\left(\frac{1}{N}\right) \tag{1.18}$$

Equation of form $y = c + xm$

It is evident that eqn 1.18 is in the form of a straight line graph. So, a plot of $\left(\frac{P}{N_s}\right)$ *versus* P should give a straight line with gradient $\frac{1}{N}$ and intercept $\frac{1}{NK}$. It is usually quite difficult to determine N_s experimentally (particularly for high surface area powders) and, therefore, more usually, changes in the mass of a

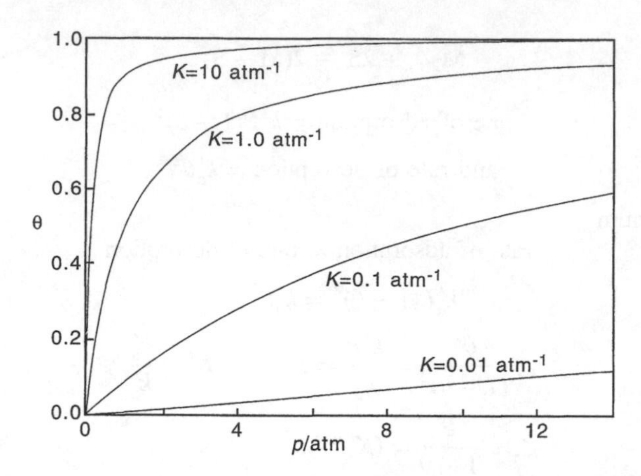

Fig. 1.1 Summary of the essential features of the Langmuir adsorption isotherm. Adapted from ref. 1.

substrate after adsorption, or changes in the volume of gas taken up by the substrate, are employed. Using eqn 1.16, equations analogous to eqn 1.18 may be formed, again of straight-line form

$$\theta = m/m_\infty \Rightarrow \left(\frac{P}{m}\right) = \frac{1}{m_\infty K} + P\left(\frac{1}{m_\infty}\right) \tag{1.19}$$

A plot of *P/m versus P* gives a straight line of gradient $1/m_\infty$ and intercept $1/m_\infty K$.

$$\theta = V/V_\infty \Rightarrow \left(\frac{P}{V}\right) = \frac{1}{V_\infty K} + P\left(\frac{1}{V_\infty}\right) \tag{1.20}$$

A plot of *P/V versus P* gives a straight line of gradient $1/V_\infty$ and intercept $1/V_\infty K$.

Equations 1.18–1.20 are the *same*, it is just the parameters used to characterize θ that are different. Moreover, each of the eqns 1.18–1.20 may also be used to calculate the total surface area of a substrate. If N, the total number of sites corresponding to a monolayer (or V_∞ the corresponding volume, or m_∞ the corresponding mass) is known and the **area of one molecule** (A_m) is known, *a priori*, then the total surface area is

$$S_A = N \times A_\mathrm{m} \tag{1.21}$$

N may be obtained from m_∞ since

$\dfrac{m_\infty}{\Omega}$ = total number of moles corresponding to one monolayer, n_m, where Ω = molar mass (g mol^{-1} if mass of adsorbate is measured in grams)

and

$$n_\mathrm{m} = N/L$$

where L = Avogadro's number (mol^{-1}).

$$\therefore\ n_\mathrm{m} = \frac{m_\infty}{\Omega} = N/L$$
$$\therefore\ N = \frac{m_\infty \times L}{\Omega} \tag{1.22}$$

Similarly, N may be obtained from V_∞ since.

$$PV_\infty = n_m RT \qquad \textit{at constant P and T}$$

$$\therefore n_m = \frac{PV_\infty}{RT} = N/L \tag{1.23}$$

$$\therefore N = \frac{PV_\infty L}{RT}$$

Although eqn 1.21 may be used to find the surface area of a solid, in order to compare the surface areas of different solids, it is more usual to express the surface area, S_A, *relative* to the mass of the substrate

$$\begin{matrix}\text{specific surface area}\\ \text{(surface area per unit mass)}\end{matrix} = S_A/(\text{mass of substrate}) \tag{1.24}$$

The specific surface area is a fundamental quantity for comparing the catalytic activity of different substances. Since heterogeneous catalysis generally occurs at the surface of solids, in order to compare 'like with like', identical surface areas should be used (for a fixed mass of catalyst a large area of one catalyst may give an 'apparently' faster rate than a small area of another, even if it is actually a poorer catalyst).

1.3 Heats of adsorption

The adsorption of a gas on a solid is an **exothermic** process. Furthermore, both the magnitude and variation as a function of coverage of the heat evolved during adsorption may reveal information concerning the type of bonding exhibited by the adsorbate to the surface and, also, evidence of **lateral interactions** between adsorbed species. Therefore, some definitions of various heats of adsorption used in surface studies will be given.

Heats of adsorption derived from calorimetric methods involve determining the heat (usually measured as a temperature rise in the solid) evolved when a known amount of gas is allowed to adsorb on to a clean surface.

The **integral** heat of adsorption at constant volume may then be expressed as

$$q_i = (Q_i/n)_V \tag{1.25}$$

where Q_i = heat evolved; n = number of moles of gas adsorbed.

If this process is repeated for various amounts of adsorbed gas, a graph may be plotted of Q_i *versus* n at constant temperature and volume (Fig. 1.2). However, it does not necessarily follow that the heat of adsorption will be independent of coverage. As molecules of adsorbate pack closer together on the surface with increasing coverage, inevitably, some lateral interactions (see section 1.13) will result, which will change the heat of adsorption. Therefore, a new function needs to be defined that takes into account possible changes in the heat of adsorption as θ varies. For the equilibrium

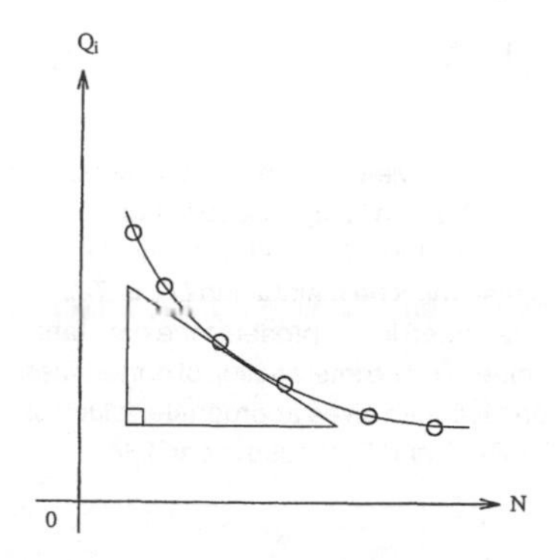

Fig. 1.2 The gradient of this graph is called the **differential heat of adsorption** $q_D = \left(\frac{\delta Q_i}{\delta n}\right)_{V,T}$

$$A_{(g)} + S \overset{K^\circ}{\rightleftharpoons} A{-}S \quad S = \text{surface site} \tag{1.26}$$

$$\Delta G^O_{AD} = -RT\log_e K^O = \Delta H^O_{AD} - T\Delta S^O_{AD} \tag{1.27}$$

or

$$\log_e K^O = -\frac{\Delta H^O_{AD}}{RT} + \frac{\Delta S^O_{AD}}{R} \tag{1.28}$$

K^{O} = thermodynamic equilibrium constant for adsorption of A on surface site S; ΔG^{O}_{AD}, ΔH^{O}_{AD} and ΔS^{O}_{AD} are the standard Gibbs free energy, enthalpy, and entropy of adsorption.

Differentiating eqn 1.28 with respect to T at constant θ

Assuming ΔH^{O}_{AD} and ΔS^{O}_{AD} are independent of T

$$\left[\frac{\partial}{\partial T}(\log_e K^{O})\right]_\theta = \frac{\Delta H^{O}_{AD}}{RT^2} \tag{1.29}$$

But rearranging the Langmuir adsorption isotherm (eqn 1.7)

$$KP = \frac{\theta}{1-\theta} \tag{1.30}$$

and taking natural logs of both sides of equation 1.30:

$$\log_e K + \log_e P = \log_e\left(\frac{\theta}{1-\theta}\right) \tag{1.31}$$

Differentiating eqn 1.31 with respect to T at constant θ, gives

$$\left[\frac{\partial}{\partial T}(\log_e K)\right]_\theta + \left[\frac{\partial}{\partial T}(\log_e P)\right]_\theta = 0$$

i.e. $$\left[\frac{\partial}{\partial T}(\log_e K)\right]_\theta = \left[\frac{\partial}{\partial T}(\log_e K^{O})\right]_\theta = -\left[\frac{\partial}{\partial T}(\log_e P)\right]_\theta \tag{1.32}$$

Putting eqn 1.32 in eqn 1.29

$$\left[\frac{\partial}{\partial T}(\log_e P)\right]_\theta = \frac{-\Delta H^{O}_{AD}}{RT^2} \tag{1.33}$$

Hence, the **isosteric (constant coverage) enthalpy of adsorption**, ΔH^{0}_{AD}, may be determined from measurements of P and T at constant coverage θ, since integration of eqn 1.33 gives

$$[\log_e(P_1/P_2)]_\theta = \frac{\Delta H^{0}_{AD}}{R}\left(\frac{1}{T_1} - \frac{1}{T_2}\right) \tag{1.34}$$

and P_1, P_2, T_1, and T_2 may be determined from the isotherm obtained at two different temperatures (Fig. 1.3).

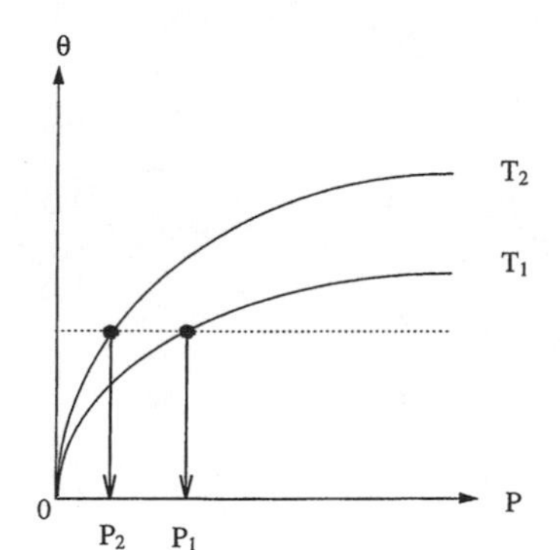

Fig. 1.3 Measurement of isosteres. Note that at all stages, equilibrium between the adsorbed layer and the gas phase must be maintained ($T_1 > T_2$). The dotted line represents the constant value of θ. The intersection of dotted lines and full lines gives appropriate values of T_1, P_1, T_2 and P_2 for use in eqn 1.34.

The relationship between isosteric adsorption enthalpy, ΔH^{0}_{AD}, the **isosteric heat**, q_{ST}, and the differential heat, q_D is

$$q_{ST} = -\Delta H^{0}_{AD} = q_D + RT \tag{1.35}$$

Isosteric heats of adsorption are often used in discussions of gas–solid equilibria. Table 1.1 lists some values of q_{ST} for nitrogen and argon.

1.4 Further discussion regarding adsorption isotherms

The BET isotherm

One major restriction in utilizing the Langmuir adsorption isotherm is that it does not allow for the fact that adsorbate film thicknesses greater than one monolayer can sometimes form. As might be expected, when **multilayer formation** does occur the heat of adsorption of an adsorbate molecule adsorbing on a bare substrate will differ significantly from the heat of

Table 1.1 Isosteric heats of adsorption for nitrogen and argon on various solid surfaces [2]

	Isosteric heat, q_{ST}, at $\theta = 0.5$		BET* surface area (m^2g^{-1})
Adsorbent	Argon (kJ mol^{-1})	Nitrogen (kJ mol^{-1})	
Carbon black	11.3	11.3	64–120
Anatase	–	14.2	11–12
Rutile	10.5	13.4	6–8
Silica	8.8	9.2	140–180
α-Alumina	10.5	13.4	1–70
γ-Alumina	8.4	11.3	180–240

* see Section 1.4.

adsorption associated with an adsorbate molecule adsorbing on to another layer of adsorbate (the strengths of adsorbate–adsorbent bonds will differ from adsorbate–adsorbate bonds). The latter case may be thought of as closely resembling **condensation** of a vapour into a liquid and the heat of adsorption should be closely similar in magnitude to the heat of vaporization of the adsorbate. By introducing a number of simplifying assumptions, the Brunauer, Emmett and Teller (BET) isotherm provides an extension of the Langmuir monolayer model to multilayer adsorption. These assumptions are as follows:

(i) Adsorption of the first adsorbate layer is assumed to take place on an array of surface sites of uniform energy (*cf.* Langmuir assumptions).

(ii) Second layer adsorption can only take place on top of first, third on top of second, fourth on top of third, etc. When $P = P_0$ (the saturated vapour pressure of the adsorbate), an infinite number of layers will form.

(iii) At equilibrium, the rates of condensation and evaporation are the same for each individual layer.

(iv) Assume that, when the number of adsorbed layers is greater than or equal to two, the equilibrium constants K^0 are equal and the corresponding value of $\Delta H^0_{AD} = -\Delta H^0_{VAP}$ (see eqn 1.29). For the first adsorbed layer, the enthalpy of adsorption is ΔH^0_{AD} as in the Langmuir case. Summation of the amount adsorbed in all layers then gives the BET equation, which expressed in linear form is

$$\frac{P}{N_s(P_0 - P)} = \frac{1}{NC} + \frac{(C-1)}{NC} \times \frac{P}{P_0} \tag{1.36a}$$

$y = c + mx$

N_s and N have the same meanings as in eqn 1.11.

$C \approx e^{(\Delta H^o_D - \Delta H^o_{VAP})/RT}$ $\qquad \Delta H^o_D = \text{enthalpy of desorption} = -\Delta H^o_{AD}$

(For a fuller account of the derivation of eq 1.36a, see [1]).

Inspection of eqn 1.16 (all symbols having their usual meanings) suggests that eqn 1.36a may also be written in terms of volumes (at constant pressure) and masses

$$\frac{P}{m(P_0 - P)} = \frac{1}{m_\infty C} + \frac{C-1}{m_\infty C} \times \frac{P}{P_0} \tag{1.36b}$$

$$\frac{P}{V(P_0 - P)} = \frac{1}{V_\infty C} + \frac{C-1}{V_\infty C} \times \frac{P}{P_0} \tag{1.36c}$$

Therefore, a plot of $\frac{P}{V(P_0-P)}$ *versus* P/P_0, for example, should yield a straight-line graph whereby the intercept on the y-axis is $1/V_\infty C$ and the slope is $(C-1)/CV_\infty$, from which both C and V_∞ may be determined. Use of eqns 1.21–1.23 with knowledge of V_∞ will give the total surface area of the substrate, S_A. Unfortunately, as with all isotherms, the range of linearity of a BET plot is usually restricted:

0.05	$<$	P/P_0	$<$	0.3
(it underestimates adsorption at low *P*)				(it overestimates adsorption at high *P*)

The shapes of various isotherms generated using eqn 1.36c are shown in Fig. 1.4 [2]. Two points concerning Fig. 1.4 should be noted. The first is the existence of a 'knee' in some of the isotherms at point B. This corresponds to the completion of the first monolayer. The second is that, as C becomes large, the isotherm resembles more and more the Langmuir adsorption isotherm with the shallow rise at B, extending further and further towards $P/P_0 = 1$, relative to the steep rise close to $P/P_0 = 0$.

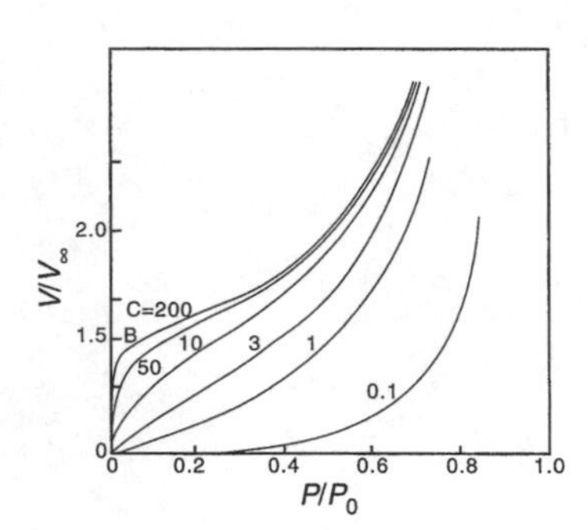

Fig. 1.4 Summary of essential features of the BET isotherm. Note that the existence of a 'knee' at point B is indicative of monolayer completion. Adapted from ref. 3.

In the limit as C $\rightarrow \infty(\Delta H^0_{AD} >> \Delta H^0_{VAP})$, eqn 1.36c reduces to

$$\frac{V}{V_\infty} = \frac{1}{1 - P/P_0} \tag{1.37}$$

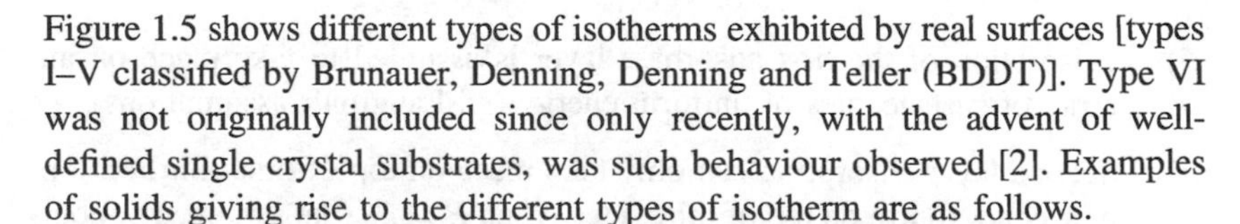

Figure 1.5 shows different types of isotherms exhibited by real surfaces [types I–V classified by Brunauer, Denning, Denning and Teller (BDDT)]. Type VI was not originally included since only recently, with the advent of well-defined single crystal substrates, was such behaviour observed [2]. Examples of solids giving rise to the different types of isotherm are as follows.

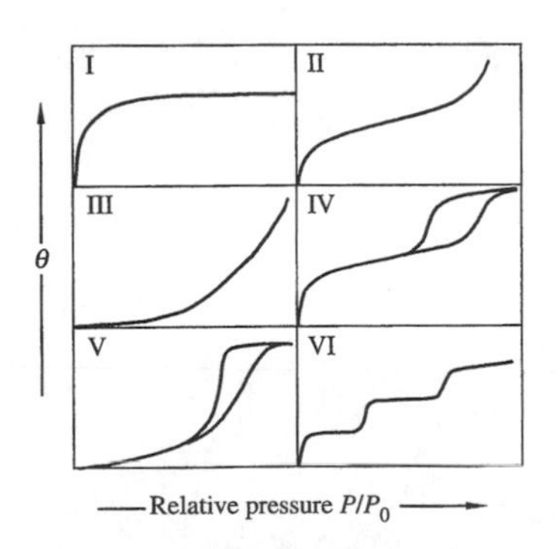

Fig. 1.5 The Brunauer, Denning, Denning, Teller (BDDT) classification of the different types of isotherms exhibited by real surfaces. See text for details.

Type I may be interpreted as closely resembling 'Langmuir' behaviour but may also correspond to filling of 'micropores' within the solid, rather than monolayer adsorption. Some activated charcoals, silica gels, and particularly 'molecular sieve' zeolites exhibit type I behaviour. Type II may be thought of as unrestricted monolayer–multilayer adsorption on a heterogeneous substrate. Hence, it may readily be interpreted within a BET isotherm to yield surface areas and thermodynamic information. Examples include nitrogen adsorption on non-porous or macroporous (pore diameter > 50 nm) powders such as carbons or oxides at 77 K. Type III behaviour is comparatively rare and, in all cases, adsorbate–adsorbent interactions are weak. H_2O adsorption on graphitized carbons or polyethylene displays a type III response. The characteristic 'hysteresis loop' of type IV systems is indicative of the presence of narrow (2–5 nm) pores that facilitate condensation. If these pores were wider, type II behaviour would be observed. Industrial adsorbents and catalyts often exhibit type IV behaviour. Type V is rather difficult to interpret and occurs rarely but type VI is readily obtained with noble gas adsorption on well-defined, uniform solids such as highly oriented pyrolytic graphite (HOPG). Each 'step' in the isotherm corresponds to the completion of the first, second, third, etc., monolayers.

A second type of deviation from ideal Langmuir adsorption arises from the assumption of independence and equivalence of adsorption sites. When this assumption is no longer valid (for example, when two or more sites with different adsorption energies exist and where energetically more favourable sites are occupied initially), it will be reflected in changes in ΔH^0_{AD} as a function of θ. In order to try to accommodate this effect, two new isotherms will be discussed briefly.

Temkin isotherm

This assumes that the adsorption enthalpy changes **linearly** with pressure by the use of two empirically determined constants, C_1 and C_2

$$\theta = C_1 \log_e(C_2 P) \qquad (1.38)$$

Freundlich isotherm

In this case, the enthalpy of adsorption is assumed to vary **logarithmically** with pressure, again with the introduction of suitable empirical constants, C_3 and C_4.

$$\theta = C_3 P^{1/C_4} \qquad (1.39)$$

The central point concerning all adsorption isotherms is that **no one isotherm can describe all behaviour over all ranges of θ and P**. None the less, over restricted ranges, each of the isotherms described above may be used to collect important data relating to the surface area of a solid and the thermodynamics of gas–solid reactions.

Again, it is emphasized, that the surface area of the solid determined using eqn 1.21 is always dependent on the assumption of close packing of the adsorbate molecules (intermolecular separations of the order of their van der Waals diameters) into a single monolayer covering the entire surface.

1.5 Bonding of adsorbate molecules to solid surfaces

Two broad classifications of adsorbate bonding may be distinguished depending on the magnitude of their enthalpies of adsorption: physisorption and chemisorption.

Physisorption

In physisorption, the bonding interaction between adsorbate and adsorbent is long range but weak and is associated with van der Waals-type interactions. Hence, bonding is characterized by a redistribution of electron density within the adsorbate and the adsorbent separately. There is negligible exchange of electrons and the value of ΔH^0_{AD} is of the order of $\Delta H^0_{condensation}$ for the adsorbate ($-\Delta H^0_{physisorption} < 35$ kJ mol^{-1}).

Examples of physisorption include the molecular adsorption of noble gases and methane. That ΔH^0_{AD} is invariably somewhat greater than $\Delta H^0_{condensation}$ is due to the fact that there is always a surface potential at the interface between two different phases, and more especially at solid–gas interfaces, whereby an 'overspill' of electron charge from the solid into the gas phase results in an imbalance of electron density on either side of the interface (see Section 2.5). The resulting surface potential generates an additional bonding interaction (ε) which becomes more significant as the **polarizability** of the adsorbate increases

α_p = polarisability of adsorbate.
E = electric field strength which is proportional to ΔV, the surface potential (Section 2.5).

$$\varepsilon = -\frac{1}{2}\alpha_p E^2 \quad (1.40)$$

Because the bonding in physisorbed systems is weak, it also tends to be reversible, in the sense that the adsorbate layer is always in equilibrium with the molecules of the gas phase. As such, physisorbed gas molecules such as argon and krypton are ideal probe molecules in determining surface area *via* Langmuir and BET isotherms.

Chemisorption

Chemisorption may usually be distinguished from physisorption on the basis of the magnitude of ΔH^0_{AD} ($-\Delta H^0_{chemisorption} > 35$ kJ mol^{-1}).

Because chemisorption is characterized by an **exchange** of electrons between the adsorbate and the adsorbent (and hence can be discussed in terms of traditional notions of covalent, ionic, and metallic bonding), spectroscopic methods can be used to confirm the nature of the surface bonding involved (see Chapter 2). Typical values of ΔH^0_{AD} are listed in Table 1.2.

The enthalpy of chemisorption depends strongly on the surface coverage of adsorbate, largely as a result of adsorbate–adsorbate lateral interactions. A good example of this effect is shown in Fig. 1.6, whereby the change in ΔH^0_{AD} for carbon monoxide adsorption on Pd(111)* as a function of CO coverage is depicted [4].

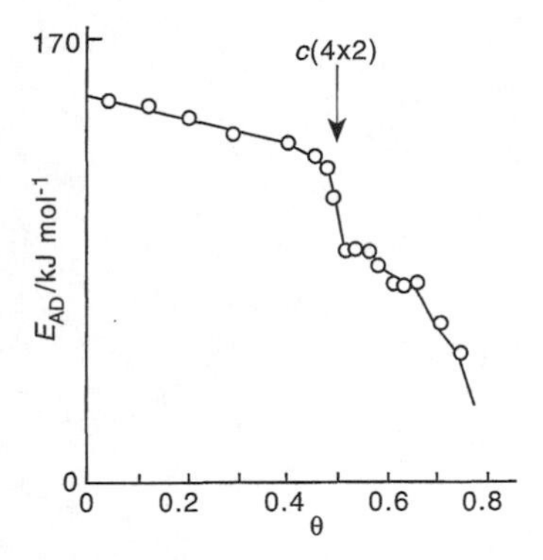

Fig. 1.6 Variation in the enthalpy of adsorption of CO on Pd(111). Adapted from ref. 4.

*The "(111)" is the Miller index plane (see section 1.8). In the above example it refers to a close-packed hexagonal arrangement of palladium surface atoms.

It is evident that as θ increases, a precipitous fall in the value of ΔH^0_{AD} occurs at exactly $\theta = 0.5$. This is associated with the formation of an ordered superlattice of surface CO molecules (see section 1.14). Further adsorption requires disruption of the ordered array and a decrease in the average separation of adsorbed CO molecules. Both effects destabilize the adsorbed layer relative to the low coverage phase and this is reflected in the decrease in ΔH^0_{AD} as chemisorption proceeds.

1.6 Kinetics of surface adsorption

Assuming Langmuir behaviour, whereby *direct* adsorption from the gas phase into a chemisorbed state occurs, the probability of a molecule being

Table 1.2 Enthalpies of adsorption for hydrogen, oxygen, and carbon monoxide on metal substrates [5]

Substrate	Gas	$-\Delta H^\circ_{AD}$/kJ mol^{-1}
Silver	Oxygen (D)	175
Platinum	Oxygen (D)	280
Tungsten	Oxygen (D)	770
Copper	Hydrogen (D)	42
Nickel	Hydrogen (D)	96
Molybdenum	Hydrogen (D)	113
Silver	Carbon monoxide (A)	27
Nickel	Carbon monoxide (A)	125
Tungsten	Carbon monoxide (D)	389

A = associative adsorption; D = dissociative adsorption.

associatively absorbed may be defined in terms of a so-called **sticking probability**, S

$$S = S_0(1 - \theta) \tag{1.41}$$

$$\text{where } S = \frac{\text{rate of adsorption of molecules by the surface}}{\text{rate of collision of molecules with the surface } (Z)}$$

and S_0 = sticking probability at $\theta = 0$. That is, the greater the number of vacant adsorption sites on the surface, the greater the probability of sticking. However, it is often observed that S is not simply a linear function of θ (Fig. 1.7) and, moreover, S can exceed the value predicted purely from Langmuir considerations! How can this be? In order to resolve this paradox, it is necessary to invoke the existence of a **precursor** state.

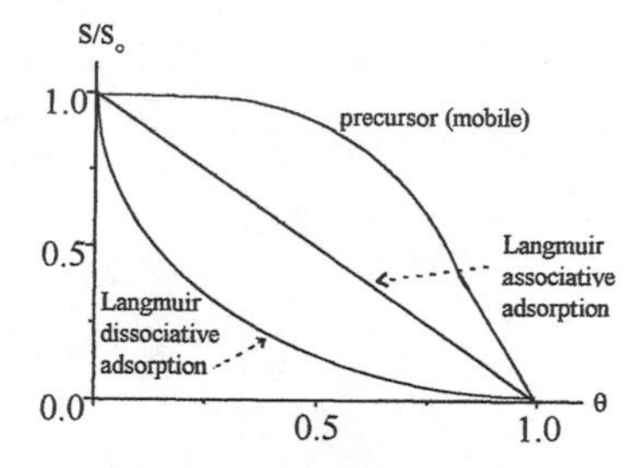

Fig. 1.7 The variation of sticking probability with surface coverage for precursor, Langmuir associative, and Langmuir dissociative adsorption.

If the adsorbate molecule collides with the surface at a filled adsorption site, it no longer necessarily rebounds back into the gas phase as postulated by Langmuir but, rather, forms a weak van der Waals-type bond to the surface and diffuses for some finite length of time (continuously losing energy as it does so) until it locates a vacant site and becomes chemisorbed. The strength of even this weak bonding interaction will be a function of whether or not the molecule finds itself 'physisorbed' to a vacant site or a filled adsorption site. In the former case, one refers to **intrinsic** precursor states, whereas the latter corresponds to **extrinsic** precursor states. There is also a need to 'dump' energy into the solid, in order for adsorption to proceed. If the energy contained within the 'hot' gas molecule is not dissipated upon impacting the surface, there will be a high probability that the 'excess' energy of the molecule will result in desorption into the gas phase. The **residence time**, τ of the precursor molecule on the surface, therefore becomes an important parameter. It is defined in terms of a pre-exponential term, τ_0, and the enthalpy of adsorption of the adsorption state

$$\tau = \tau_0 e^{-\Delta H^{\circ}_{AD}/RT} \tag{1.42}$$

If the weak adsorbate–substrate bond formed as a consequence of entering the physisorbed state is assumed to have a vibrational lifetime, τ_0 of 10^{-13} seconds (typical of molecular vibrations), the residence time may be estimated.

Equation 1.42 is evidently a simple Arrhenius-type expression, but the central point, illustrated in Table 1.3, is that the residence time a molecule spends on a surface critically depends on both the magnitude of the physisorption/chemisorption potential energy well into which it adsorbs and the substrate temperature.

Table 1.3 Values of the residence time, τ, assuming $\tau_0 = 10^{-13}$ s calculated using eqn 1.42

$\Delta H^{\circ}_{physisorption}$	$= -30$ kJ mol^{-1}
$T = 100$ K	$\tau = 469$ s
$T = 200$ K	$\tau = 7 \times 10^{-6}$ s
$T = 300$ K	$\tau = 2 \times 10^{-8}$ s
$\Delta H^{\circ}_{chemisorption}$	$= -100$ kJ mol^{-1}
$T = 300$ K	$\tau = 25833$ s
$T = 400$ K	$\tau = 1$ s
$T = 500$ K	$\tau = 3 \times 10^{-3}$ s

The longer the molecule resides at the surface, the more probable is the process of energy exchange (thermalization) with the surface. In general, the deeper the potential energy well corresponding to the bond between precursor and surface, the longer the residence time. The rate at which molecules colliding with a surface can lose their energy (and become adsorbed) is measured in terms of the **thermal** accommodation coefficient, α

$$\alpha = \frac{T_f - T_i}{T_s - T_i} \tag{1.43}$$

where T_i = initial temperature of molecule in the gas phase; T_f = final temperature of molecule in the gas phase, after collision with the surface; and T_s = temperature of surface. Hence, when $T_i = T_f$ (no exchange of energy

Table 1.4 Thermal accommodation coefficients for high energies of the gas molecules $\alpha(\infty)$[6].

Substrate	Mass	H_2 ($M = 2$)	N_2 ($M = 28$)	Xe ($M = 131$)
C	12	0.29	0.5	0.18
Si	28	0.15	0.6	0.35
Fe	56	0.08	0.53	0.50
Rh	103	0.04	0.40	0.59
Pt	195	0.02	0.26	0.58

between gas molecule and surface), $\alpha = 0$ and gas molecules are referred to as being **elastically** scattered. When $T_f = T_s$ (thermalization of gas with surface), $\alpha = 1$ and complete accommodation of the adsorbate molecule takes place. As long as the initial kinetic energy of the molecule in the gas phase ($\frac{3}{2}kT_i = E_i$) is less than $E_{precursor}$ (the depth of the physisorption well), the molecule will gradually lose energy and become accommodated. However, when E_i is similar to $E_{precursor}$, energy exchange becomes less efficient and α falls (α passes through a minimum when $E_i = E_{precursor}$). As E_i exceeds $E_{precursor}$, α starts to increase again up to the **classical limit**, whereby at very high gas temperatures ($T \to \infty$) the energy transfer may be calculated in terms of a simple inelastic collision between one particle (the adsorbate) and another (the surface atom), to give

$$\alpha(\infty) = \frac{2.4\,\mu}{(1+\mu)^2} \qquad \mu = \frac{\text{mass of adsorbate atom}}{\text{mass of surface atom}} \tag{1.44}$$

Energy transfer in the classical limit of high E_i is most efficient when the adsorbate and surface atoms are of similar mass ($\mu = 1$). This may be seen from Table 1.4.

Thus, in summary, a mechanism that accounts for the possibility of precursor-mediated adsorption may be described by

$$A_{(g)} \underset{k_d}{\overset{k_t}{\rightleftharpoons}} A_{(p)} \overset{k_a}{\rightarrow} A_{(ads)} \tag{1.45}$$

where k_t = rate constant for **trapping** of molecule into precursor state, p; k_d = rate constant for desorption; and k_a = rate constant for adsorption from precursor state.

The probability of a molecule being adsorbed into a precursor state is defined by the trapping coefficient, β

$$\beta = k_t/Z \tag{1.46}$$

Z = rate of collision of molecules with surface, and the relationship between **trapping** and **sticking** may be expressed as

$$S_0 = \frac{\beta}{1 + k_d/k_a} \tag{1.47}$$

The presence or absence of precursor states can have a profound effect on the rates of adsorption and desorption. For example, although favoured thermodynamically, the rate of dissociative chemisorption of a diatomic

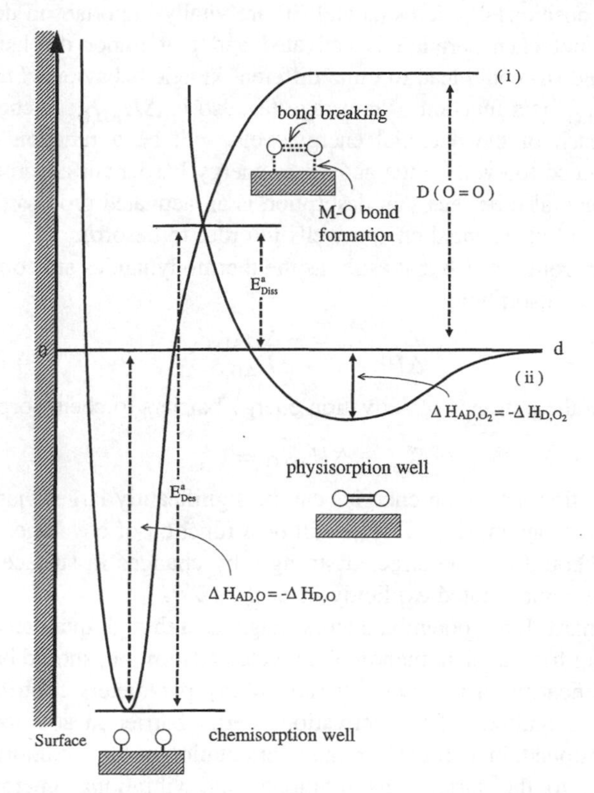

Fig. 1.8 One-dimensional potential energy diagram for dissociative adsorption.

molecule, such as nitrogen on W (110), is observed experimentally to proceed very slowly. Such slowness in the rate of a chemical reaction strongly implies the presence of an activation energy barrier to dissociation. One of the first attempts to explain such behaviour was the one-dimensional potential energy diagram for dissociative adsorption proposed by Lennard–Jones (Fig. 1.8) which also incorporates the possibility of precursor-mediated kinetics.

Figure 1.8 shows a one-dimensional plot of potential energy *versus* reaction coordinate for the reaction

$$O_{2(g)} \rightarrow O_{2(precursor)} \rightarrow 2O_{(ads)} \tag{1.48}$$

Two curves are displayed associated with: (a) the interaction of the diatomic molecule (oxygen in this case) with the surface [curve (ii)]; and (b) the interaction of oxygen atoms with the surface [curve (i)]. Adsorption into the precursor state [curve (ii)] is a non-activated process (no activation energy barrier) and $\Delta H_{AD,O_2} = -\Delta H_{D,O_2}$. However, in order for the oxygen molecule to enter into a chemisorbed atomic state, it must overcome the activation energy barrier, E^a_{Diss} formed at the crossing of curves (i) and (ii). The magnitude of this barrier to dissociation (*via* surface adsorption) should be contrasted, however, with the much larger gas phase dissociation energy D(O=O). Clearly, because of the disparity between D(O=O) and E^a_{Diss}, dissociation may be facilitated by the surface with respect to the gas phase. Hence, **activated** adsorption corresponds to the point of crossing of curves (i) and (ii) lying above the energy zero reference line. **Non-activated** adsorption is said to occur when the crossing of curves (i) and (ii) lies below the energy zero reference line. It is evident that the relative positions of curves (i) and (ii) are vitally important in determining whether or not chemisorption is activated and that minor displacements of curves (i) and (ii) could lead to quite different kinetic behaviour. Furthermore, since $\Delta H^0_{AD,O_2}$ is a function of coverage (as also is $\Delta H^0_{AD,O}$; Section 1.3) the depths of each of the potential energy wells will be a function of surface coverage and so too will be the activation energy barrier for adsorption. From Fig. 1.8, it can also be seen that desorption is an activated process (adsorbates must climb out of potential energy wells in order to desorb).

It is clear from Fig. 1.8 that as far as the thermodynamics are concerned for non-activated adsorption:

$$\Delta H_{D,O_2} = -\Delta H_{AD,O_2} \tag{1.49}$$

and

$$\Delta H_{D,O} = -\Delta H_{AD,O} \tag{1.50}$$

However, in the presence of activation energy barriers to chemisorption, from Fig. 1.8

$$E^a_{Des} + \Delta H_{AD,O} = E^a_{Diss} \tag{1.51}$$

Thus, desorption activation energies can be significantly larger than those for adsorption and, again, since ΔH_{AD} will be a function of coverage, the rate of desorption should also be affected strongly by changes in surface coverage. This will be demonstrated explicitly in Section 2.7.

The Lennard–Jones potential energy diagram, although qualitatively useful in accounting for the phenomenon of activated adsorption, should be seen as a gross simplification of the real situation. Many parameters contribute to the variation in magnitude of the activation energy barrier in addition to those already mentioned, including the relative orientation of the incoming molecule with respect to the surface, its rotational and vibrational energy, and its position above a particular site (directly above or between surface atoms) at the point of impact. To illustrate this point, consider the dissociative adsorption of hydrogen on copper. This process is activated and S_0 is very low. Hence, in order to achieve adsorption, one could simply increase the translational energy (E_t) of the gas phase hydrogen molecules (increase their temperature) and eventually, at sufficiently high values of E_t, hydrogen atoms will begin to adsorb on to the copper surface. Once they have sufficient energy to surmount the activation energy barrier. However, the key step in breaking the H–H bond is to stretch it first (bond strength decreases as overlap between hydrogen atomic s orbitals diminishes, until, eventually, the molecule dissociates). One way of achieving this would be to excite the molecule **vibrationally** (average H–H bond length increases), and, in fact, the vibrationally excited state of hydrogen requires less translational energy for dissociation than does a molecule in the ground state. Increased rotational energy also helps to dissociate the molecule, at least as long as the rotation is in the plane of the surface, as illustrated in Fig. 1.9.

Hence, to account fully for the real **dynamics** of surface adsorption, a multidimensional potential energy surface is required inclusive of positional, translational, rotational, and vibrational degrees of freedom. In Section 2.8, experimental probes of the dynamics of surface adsorption will be outlined more fully.

To conclude this section, it is instructive to view the adsorption process simultaneously in terms of potential energy diagrams and changes in filling of molecular orbitals on the molecule and electronic bands of the solid.

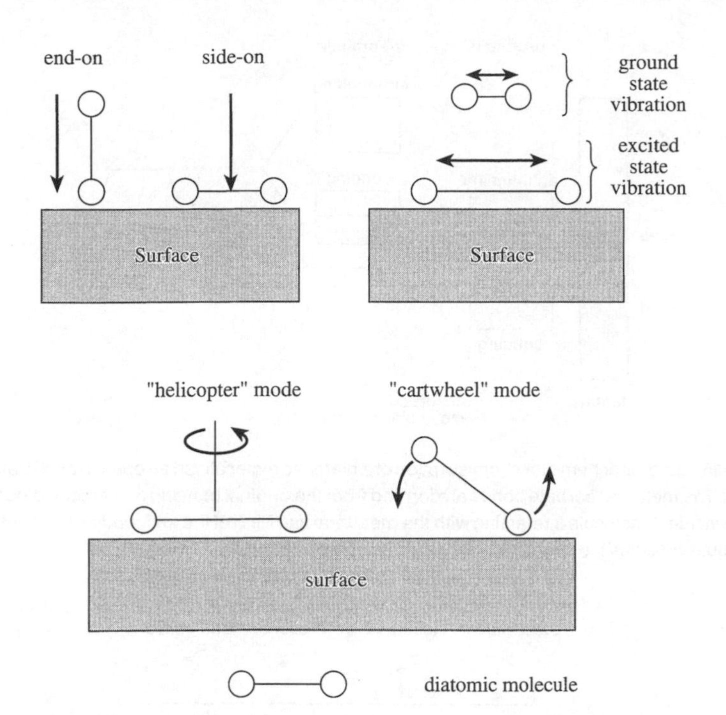

Fig. 1.9 ORIENTATION of molecule will affect rate of dissociation. May be interpreted as differences in extent of overlap between antibonding orbital on molecule and filled electronic states on solid. **VIBRATIONAL** state of molecule will affect rate of dissociation. Bond length in diatomic molecule is longer on average in excited vibrational state. **ROTATIONAL** excitation also leads to increase in molecular bond length. However, poor orbital overlap between molecule and surface is observed in the 'cartwheel' mode relative to the 'helicopter' mode. Therefore, a lower probability of sticking is encountered where the molecule adopts a 'cartwheel' configuration with respect to the surface.

Figure 1.10 shows, schematically, how the molecular orbitals of gaseous dihydrogen are transformed as they interact with a metal surface. Two effects should be noted. The first is a broadening in the energy range of the individual molecular orbitals of dihydrogen relative to the free molecule. This occurs because of gradual mixing of electron wavefunctions on the metal and the molecule. The second is that two new sets of molecular orbitals between the adsorbed molecule and the metal (σ bonding and σ^* antibonding) develop. In addition, electron transfer from the metal takes place into σ^* orbitals of the dihydrogen until the highest occupied electron state on the molecule is equal in energy to the electrons at the Fermi energy E_f of the metal. Depending on the extent of electron transfer from the metal to dihydrogen (ostensibly now in a precursor state), the H–H bond will weaken until the bonding component of the bond energy is outweighed by the antibonding component and dissociation into adsorbed atoms occurs. The extent of electron transfer will depend on the relative positions of E_f and the vacant σ^* orbitals on the precursor. Those metals that can readily transfer charge to the σ^* orbitals of dihydrogen will effect dissociation, whereas those that do not (e.g. copper) will tend to exhibit a low value of S_0 (see also Section 2.6).

The valence electrons in a metal occupy an effective continuum of energy states called a band. The energy state corresponding to the highest energy electrons within this band is called the Fermi level.

Figure 1.11 shows the corresponding orbital diagrams for the complete dissociation of a diatomic molecule.

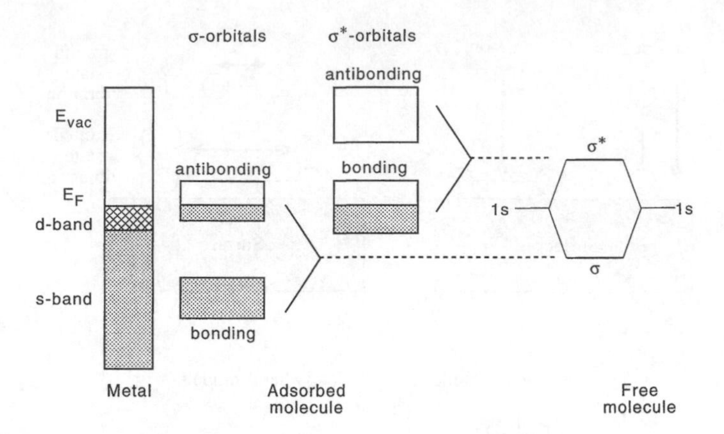

Fig. 1.10 Orbital scheme for chemisorption of a diatomic molecule on an open d-band transition metal. The metal–adsorbate bonds are formed from the original bonding and antibonding levels of the incident molecule interacting with the metal. Partial filling of the antibonding levels of the molecule weakens the intramolecular bond. Adapted from ref. 6.

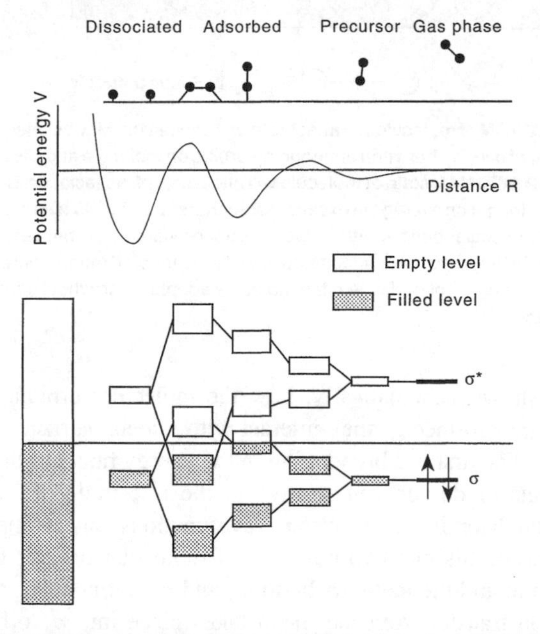

Fig. 1.11 Potential energy diagram for the approach of a diatomic molecule from the gas phase to the surface along with the corresponding orbital diagrams. Adapted from ref. 6.

As the diatomic molecule approaches closer to the surface (gas $\rightarrow$ precursor $\rightarrow$ molecular adsorption) the σ^* and σ orbitals on the free molecule are seen to split into two new energy levels (as in Fig. 1.10), the extent of splitting being a function of the molecule–solid distance.

Eventually, what was the σ^* orbital on the free molecule falls below the energy of the Fermi electrons, E_f, and becomes partially filled with electrons transferred from the metal (electrons on the metal try to lower their energy by moving to vacant orbitals of lower energy on the molecule). The filling of σ^* orbitals by electrons from the metal is referred to as **back-donation**. When the diatomic molecule dissociates, forming chemisorbed atoms, the orbitals of the molecularly adsorbed state collapse to give two simple M–H σ and M–H σ^* molecular orbitals (M = metal). Photoelectron spectroscopy can be used to monitor the changes in the distribution of electrons in molecular orbitals described above (see Section 2.6).

1.7 The study of well-defined surfaces

Up to now, our picture of what a solid surface looks like at the level of atoms and molecules has remained rather vague. It is evident from the previous discussion (for example sticking probabilities) that the chemical and physical properties of the surface will depend ultimately upon its electronic structure. This, in turn, will be a function of the nature of the atoms comprising the surface and their spatial distribution. Recently, with the advent of scanning probe microscopies (see Section 2.4), it has become possible to image individual surface atoms. Invariably, surfaces are found to consist of a mixture of flat regions (called **terraces**) and defects (**steps, kinks**, and **point defects**; see Fig. 1.12). Since the local distribution of atoms around each of these individual **surface sites** is different, one should expect their electronic properties to be distinguishable as well. Indeed, this is found to be the case. Hence, each surface site will exhibit its own singular surface chemistry and physical response. This makes the experimental and theoretical investigation of 'rough' solid surfaces (99.99% of cases) at the atomic level an almost intractable problem, since any information gathered must, by definition, contain contributions from a myriad of different combinations of surface sites and surface compositions. Therefore, in order to ensure that an experimentalist in one laboratory obtains the same *reproducible* result on a particular type of solid surface as another, it is necessary to define precisely the chemical and structural state of the substrate under investigation. This involves first considering the most simplified systems of all: surfaces containing a high ratio of terrace to defect sites (normally referred to as 'flat') and consisting of just one type of atom. Although clearly not representative of the types of solid surfaces normally encountered, this approach resolves all of the myriad structure/composition combinations down to just a single type of site. Greater complexity can be introduced into the system by adding *controlled* amounts of surface defects or coverages of chemically distinguishable adsorbates. Such surfaces are referred to as being *well-defined*. Over the last 30 years, a large database has been established in which the properties of well-defined surfaces (metals, oxides, semiconductors, insulators) have been intensively investigated. Because of this, at the time of writing (1997), a good understanding of solid–gas and solid–liquid phenomena has been attained based on the **surface science approach**. In what follows, the experimental prerequisites necessary to investigate well-defined systems will be outlined. In Chapter 2, the type of information that may be obtained as a consequence of this approach will be developed further.

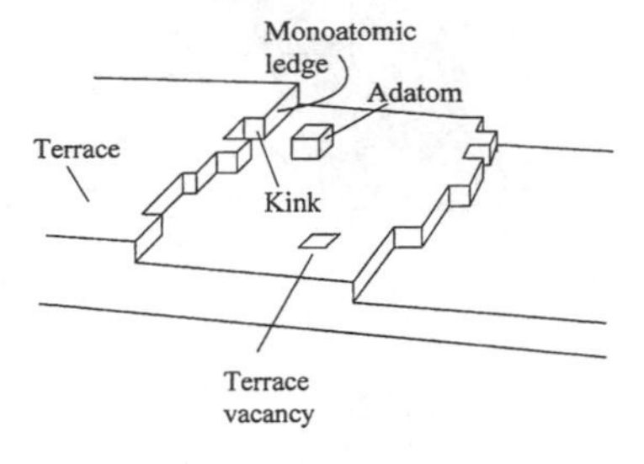

Fig. 1.12 Schematic representation of step, kink, terrace, adatom, and vacancy model of surface sites.

1.8 Single crystal surfaces

To ensure that experiments are performed on well-defined surfaces, single crystals are invariably used as substrates. This is because a given cut through such a crystal will expose a particular crystal plane. By definition, such planes will contain atoms in a *limited* number of well-defined sites. The number and/or geometrical arrangement of adsorption sites in the surface may be varied systematically by simply cutting the single crystal in different directions to expose different crystallographic planes. To decide on a particular direction of cut, the Laue (X-ray back-scattering) technique is used to align diffracted X-ray beams (associated with the desired plane) originating from the single crystal, such that the correct orientation of the cut (*via* spark erosion or grinding) is maintained at all times during the preparation procedure. Surfaces produced in this way may be either atomically **flat** (consisting solely of large terraces) or **vicinal** (consisting of short atomically flat terraces separated by atomic steps).

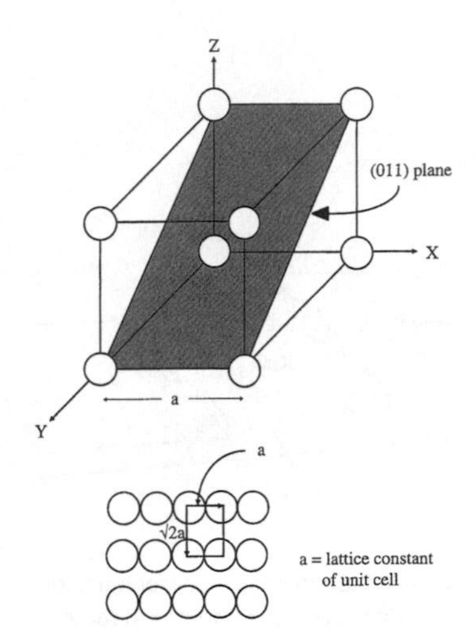

Fig. 1.13 The (011) plane of a simple cubic lattice.

The **Miller index** is used to denote a given plane. It consists of three integers (x, y, z) for materials adopting bulk cubic [simple cubic (sc), face-centred cubic (fcc), body-centred cubic (bcc)] or four integers (w, x, y, z) for those materials exhibiting hexagonal close-packed (hcp) structures. There are other less common crystallographic forms [7] which may also be identified using Miller indices but which will not be the subject of this introductory text. As an example of defining the Miller index of a plane, consider Fig. 1.13 in which three mutually perpendicular Cartesian axes, labelled x, y, and z, define a simple cubic lattice of lattice constant 'a'. The crystallographic plane to be labelled is indicated by the shaded area. To determine the Miller index of the plane, the following operations are performed.

(i) Decide where the plane intercepts the x-, y-, and z-axes in multiples of the unit-cell dimension, a

$$\text{intercepts} = \infty, 1, 1$$

Note that the intercept of the plane with the x-axis is at infinity since the plane runs parallel to it. If a negative intercept happens to result, this is indicated by placing a bar above the index. Hence, $(\infty, 1, -1)$ would become $(\infty, 1, \bar{1})$.

(ii) Take the reciprocal values of these intercepts.

$$\text{reciprocals} = (1/\infty, 1/1, 1/1) = (0, 1, 1)$$

(iii) If fractions result from step (ii), reduce the three indices to the equivalent ratio of whole numbers. For example, if step (ii) gave (1/3, 2/3, 1), the corrected Miller index would be denoted (1, 2, 3).

Crystallographic directions by convention are enclosed in square brackets e.g. [110]. However for Miller Index planes, less rigorous conventions apply and both curly {} and round () brackets are used. In addition, the commas separating the integers within the bracket are often deleted for simplicity.

Figures 1.14 and 1.15 show the principal low index planes (Miller indices containing only zeros and ones) of fcc and bcc crystals, respectively. In addition, the important crystallographic directions within each plane (the three indices in square brackets), the interatomic separations in terms of the bulk lattice constant 'a', and the *primitive* surface unit mesh (broken lines) associated with each plane are shown. By primitive, is meant the smallest possible repetitive unit cell necessary to generate the surface lattice. Hence, from Fig. 1.15, the primitive unit cell of a bcc (1, 1, 0) surface is a rhombus. However, one could equally well take as the unit cell the centred rectangle (also indicated in Fig. 1.15). Although capable of generating the surface lattice, the centred rectangle unit cell is not the smallest unit and, therefore, is

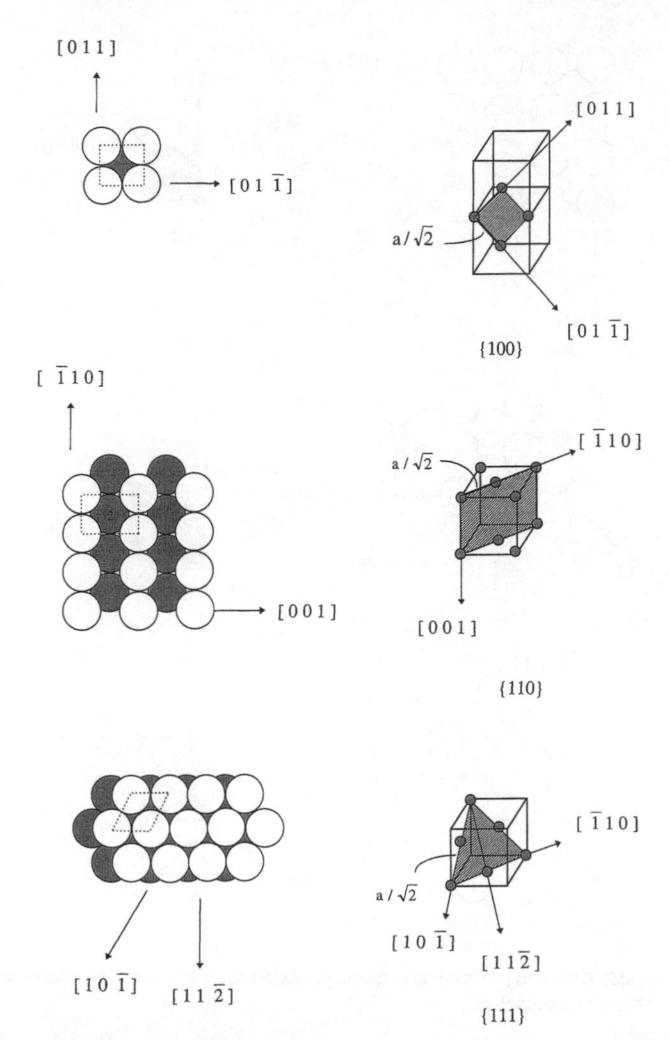

Fig. 1.14 The (100), (110), and (111) planes of fcc crystals. Second layer atoms in the (110) plane are labelled '2'. For clarity, atoms in the centre of the cube faces have been left out unless contained within the cut plane.

not considered primitive. In fact, there are only *five* possible surface unit meshes, or Bravais lattices, in two-dimensional systems (four primitive and one centred). These are described in Fig. 1.16.

As mentioned previously, some elements crystallize in non-cubic forms, and one of the most common of these is the hexagonal close-packed structure consisting of close-packed hexagonal layers placed one on top of the other with a repeat stacking pattern, ABAB...

In order to preserve the symmetry of hcp systems, four axes need to be defined, the first three (w, x, y) oriented at 120° to each other, located in the plane of the hexagonally close-packed layer, and the fourth (z) orthogonal to the close-packed plane. The simplest plane of hcp materials is formed by cutting perpendicular to the z-axis and parallel to the (w, x, y)-axes. This plane is shown in Fig. 1.17 and is termed the **basal** plane. Also shown in Fig. 1.17 is the $(10\bar{1}0)$ hcp plane.

Planes of high Miller index (containing indices > 1) are not flat on an atomic scale but consist of narrow, low-index planes separated by steps,

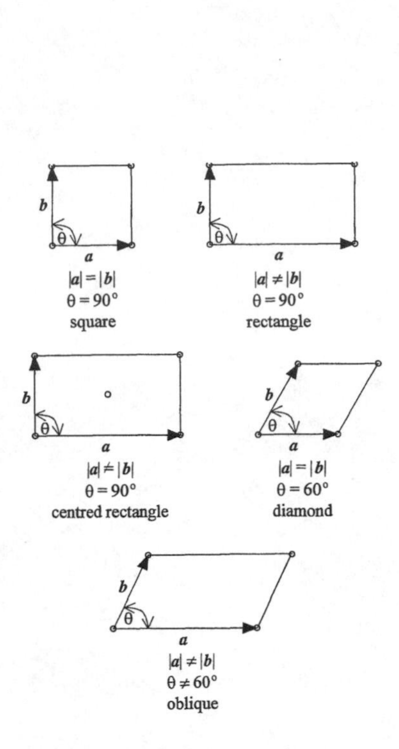

Fig. 1.16 The five Bravais surface lattices.

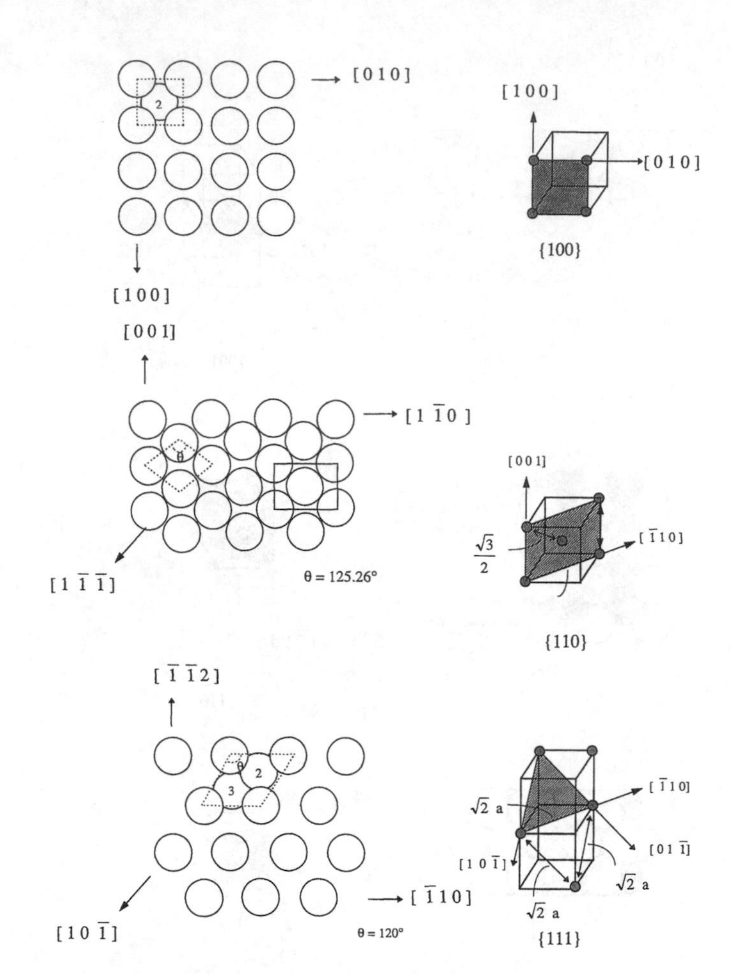

Fig. 1.15 The (100), (110), and (111) planes of bcc crystals. Second and third layer atoms are indicated by a '2' and '3', respectively.

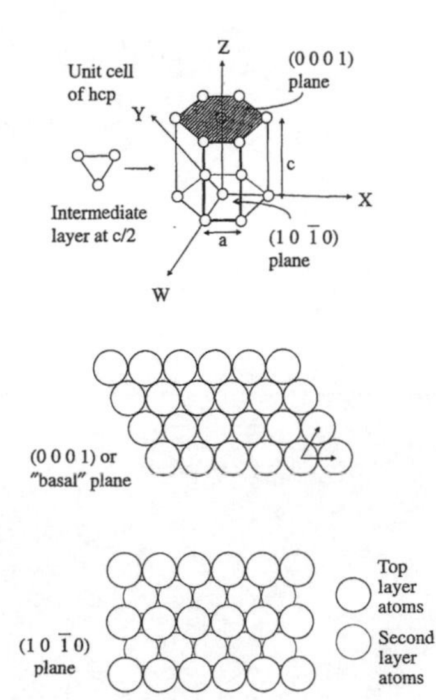

Fig. 1.17 The (0001) basal plane and (10$\bar{1}$0) planes of a hcp crystal. The unit cell of an hcp crystal is also displayed (top).

usually one atom high. Such 'staircase' structures are more conveniently referred to using a 'microfacet' notation

$$n(x, y, z) \times (u, v, w)$$

where n is the average number of atoms in the terrace, (x, y, z) is the Miller index of the terrace and (u, v, w) is the Miller index of the step. Hence, an fcc (11, 1, 1) surface may alternatively be labelled 6(100)×(111), i.e. a series of six-atom wide (100) terraces separated by (100)×(111) steps. Similarly, an fcc (331) plane may be referred to as 3(111)×(111), i.e. three-atom wide (111) terraces separated by (111)×(111) steps. Examples of various high Miller index planes and their alternative microfacet notation are shown in Fig. 1.18. In fact, all crystal planes may be described by various combinations of the three low-index planes. This idea is illustrated in Fig. 1.19 where the **stereographic triangle** of the fcc system is described. It should be thought of as analogous to a three-component phase diagram [the three components being the (111), (100), and (110) low-index planes]. The further one travels along a particular direction away from the three poles (the three corners of the stereographic triangle), the less is the proportion of a particular terrace in a

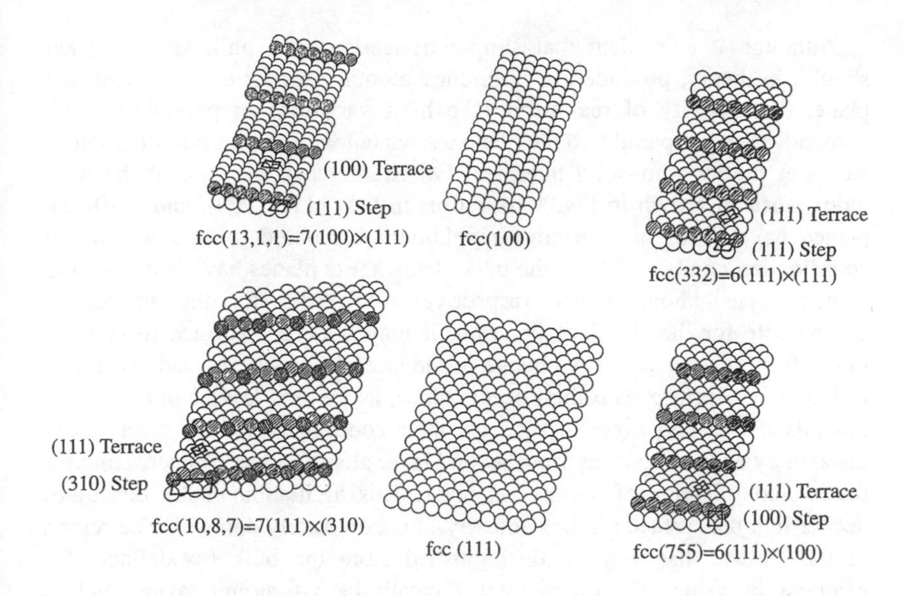

Fig. 1.18 Examples of various high Miller index planes for a fcc crystal. The microfacet notation for the stepped surfaces is also shown.

given plane. Hence, from Fig. 1.19, travelling from the (111) pole along the $[01\bar{1}]$ direction to the (100) pole, the size of the (111) terraces decreases and (eventually) the size of the (100) terrace becomes larger. At $(311) = 2(100) \times (111) = 2(111) \times (100)$ it should be noted that steps and terraces become indistinguishable! This point is referred to as the **turning point** of the zone. Points within the stereographic triangle (away from the edges) correspond to **kinked** high Miller index planes (surfaces containing regular non-linear steps).

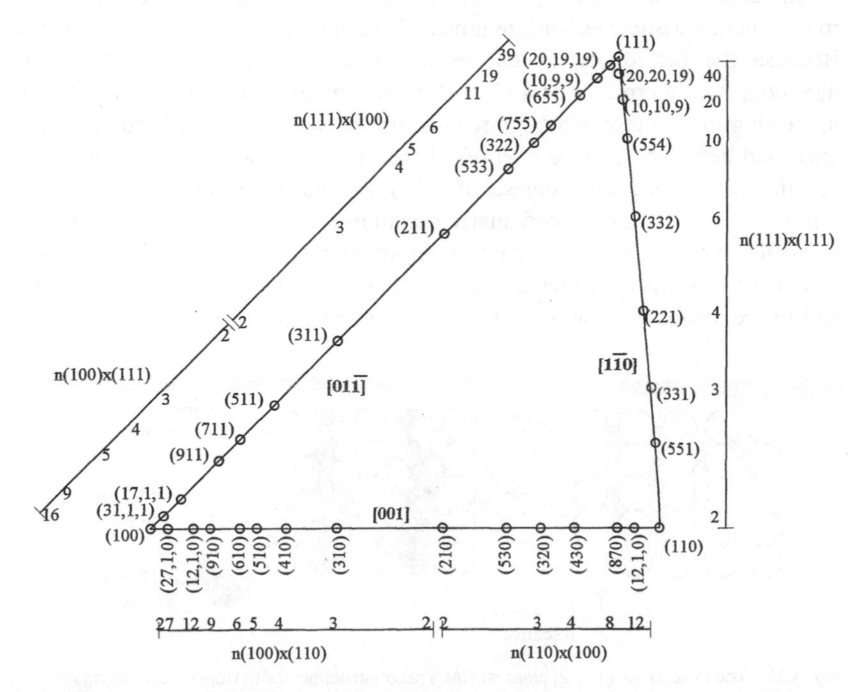

Fig. 1.19 The stereographic triangle for the fcc crystal system. The value of n for each of the Miller indices listed is also shown.

From Fig. 1.19 can you decide the Miller index of the turning point travelling from the (100) pole to the (110) pole?

Although it is evident that simple truncation of a bulk single crystal should, in theory, produce a well-defined atomic periodicity in the exposed plane, the majority of real surfaces exhibit variations in periodicity both perpendicular and parallel to the plane associated with loss of coordination of atoms at the vacuum–solid interface. For example, in the case of the low-index surfaces shown in Fig. 1.14, atoms in the (111), (100), and (110) fcc planes have 9, 8, or 7 nearest neighbours compared with their normal coordination number of 12 in the bulk. Thus, these planes have lost 3, 4, and 5 nearest neighbour bonds, respectively. In order for the surface to compensate for the 'loss' of bonding, it may undergo 'surface relaxation' in the form of an oscillatory change in the interplanar spacing Δd, as shown in Fig. 1.20. Surface relaxation occurs since, as the first layer atoms contract towards the second layer to increase their coordination, third layer atoms respond by expanding away from second layer atoms, hence compensating for the overcoordination of the second layer. This oscillation in Δd penetrates deeper into the surface until, eventually, it is completely damped. The region of the surface that may be distinguished from the bulk (as defined, for example, by values of $|\Delta d| > 0$ will typically be 5–6 atomic layers thick at most, and is termed the **selvedge**. Surface relaxation is, in general, largest for low atomic density (more open), high energy surfaces. The surface energies reflect the degree of coordination of atoms in the surface. Hence, for fcc metals, the surface energy decreases in the order: (110) > (100) > (111), and for bcc surfaces: (111) > (100) > (110).

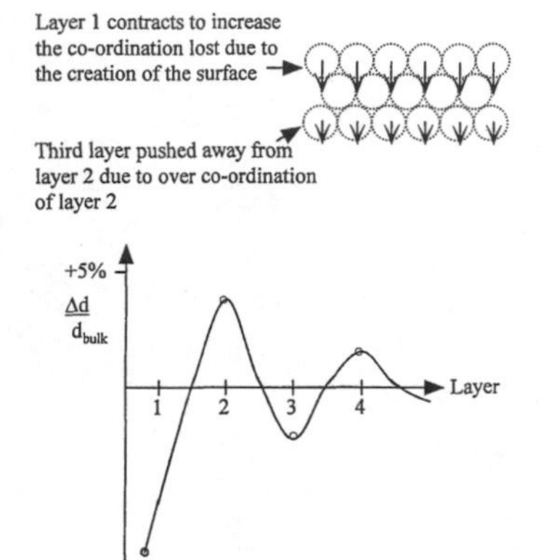

Fig. 1.20 Surface relaxation in the form of an oscillatory change in the interlayer spacing, Δd, as a function of layer thickness.

If the surface energy is sufficiently large, not only will surface relaxation occur but also gross restructuring of the surface plane, usually to enhance the coordination of surface atoms and, hence, to achieve a lower surface energy. When simple truncation of the bulk crystal does *not* lead to the 'expected' surface periodicity, the effect is termed **surface reconstruction**. Figure 1.21 illustrates the (110) plane of gold which reconstructs when clean to a 'missing-row' structure associated with removal of alternate close-packed atomic rows. Because the periodicity of the reconstructed surface is doubled in one direction, it is referred to as a (1 × 2) reconstruction (see Section 1.15). It is interesting to note that, although reconstruction increases the effective surface area (and hence the surface energy) of the gold, this increase is offset by the fact that the surface now consists of (111) microfacets of low surface energy. Surface reconstruction is particularly common in semiconductors, which tend to exhibit more localized covalent bonding. For example, Si(100) undergoes row pairing between surface atoms associated with the formation of π bonds and hence also gives rise to a (1 × 2) reconstruction.

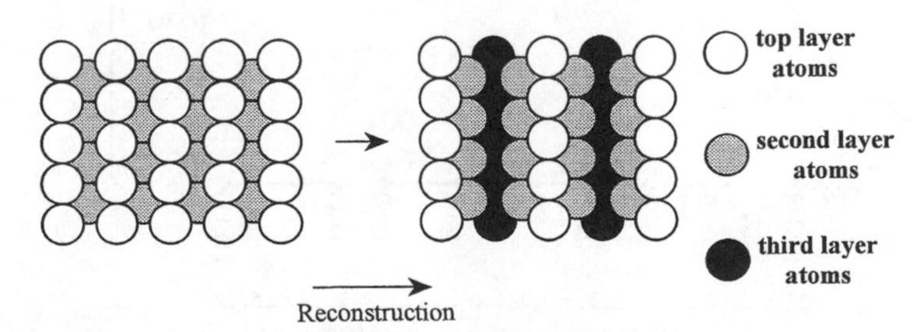

Fig. 1.21 The (1 × 1) → (1 × 2) clean surface reconstruction of Au(110)—the 'missing row' model.

1.9 Preparation and maintenance of atomically clean surfaces

Two fundamental problems associated with the study of solid surfaces are, first, how does one prepare a *clean* solid surface and, second, how, once clean, does one keep the surface clean and well-defined over the period of an experiment (tens of minutes to perhaps several hours)? The latter problem may be understood if it is realized that the rate of contamination will depend upon the rate at which gaseous molecules collide with the surface. From consideration of the kinetic theory of gases, the rate of surface bombardment by molecules (Z) is given by

$$Z = \frac{p}{(2\pi mkT)^{1/2}} \text{ cm}^{-2}\text{ s}^{-1} \tag{1.52}$$

where p = ambient pressure, in Ncm^{-2}; m = molecular mass, in kg molecule^{-1}; T = absolute temperature, in K; and, k = Boltzmann constant, in J K^{-1}.

Of course, the rate of surface contamination will also depend upon the sticking probability $S(\theta)$ (the gas molecule may collide with the surface but it may not necessarily stick!). Assuming a worst case scenario of $S(\theta) = 1$, it is instructive to use eqn 1.52 to estimate the coverage of CO (a typical gaseous contaminant) produced at 300 K and ambient pressures of 10^{-6} Torr and 10^{-10} Torr, respectively (1 Torr = 1.333×10^{-2} Ncm^{-2})

At 10^{-6} Torr

$$Z = \frac{(1.333 \times 10^{-2} \times 10^{-6})}{[2 \times \pi \times (\frac{28}{1000 \times 6.02 \times 10^{23}}) \times (1.38 \times 10^{-23}) \times 300]^{1/2}} \text{cm}^{-2}\text{ s}^{-1}$$

$$3.82 \times 10^{14} \text{ cm}^{-2}\text{ s}^{-1}$$

Assuming an atomic density of 10^{15} cm^{-2} (typical of most surfaces), this means that the rate of surface contamination by CO $[S(\theta) = 1]$ at 300 K would be

$$\frac{3.82 \times 10^{14} \text{ cm}^{-2}\text{ s}^{-1}}{10^{15}\text{ cm}^{-2}\text{ per monolayer}} = 0.382 \text{ monolayers s}^{-1}$$

∴ Time taken to adsorb one monolayer at 10^{-6} Torr = 1/0.382 = 2.6 s

A similar calculation for $p = 10^{-10}$ Torr gives 26 178 seconds or 7.3 hours!

Clearly, from the calculations above, in order to maintain a clean surface, the ambient pressure needs to be reduced to as small a value as possible. 'Ultra-high vacuum', or UHV, signifies that the ambient pressure has been reduced to below 10^{-9} Torr. *The achievement of UHV is an essential experimental condition which needs to be met in order to study reactive solid–gas interfaces.*

A unit used generally for defining gas exposures is the Langmuir (L). This is defined as an exposure of 10^{-6} Torr for 1 second, or 10^{-7} Torr for 10 seconds, or 10^{-8} Torr for 100 seconds, etc.

But how does one produce a clean surface? The answer to this question depends largely upon the type of solid to be studied. Some layered compounds, such as graphite, mica, and a number of semiconductor materials, may simply be cleaved in UHV to produce clean, well-ordered surfaces. If a well-defined surface structure is not important, mechanical scraping of the outermost contamination layer will also suffice. In the case of many semiconductors, atomically clean surfaces grown by depositing gaseous atoms on to a suitable substrate may be 'capped' with a thin protective layer. Such samples may be transported through air with the 'capping layer' protecting the surface. Being made from a volatile material, the cap may be desorbed subsequently in UHV by heating, leaving behind the clean semiconductor surface for analysis. Cleaning by chemical reaction is often useful in removing carbonaceous layers through heating of the sample in an oxygen-rich environment to produce gaseous CO and CO_2. The excess adsorbed oxygen atoms may be removed as gaseous H_2O by subsequent gentle heating of the sample in hydrogen. Supported metal catalysts are often prepared in this manner.

By far the most common method of surface cleaning, however, is 'argon ion bombardment'. This method consists of the physical removal of surface material, typically at a rate of several monolayers per minute, by bombardment of the surface with a beam of high energy argon ions (100–3000 eV). Upon striking the surface, energy transfer from the argon ions to the substrate causes surface atoms to break their bonds with the substrate and desorb into the vacuum (this is called sputtering). Because of the violent nature of the process, argon ion bombardment leaves a pitted and rough surface. Therefore, annealing of the substrate to temperatures below or close to the substrate melting point is essential, both to facilitate surface diffusion (and hence the removal of defect sites so that a flat, clean surface is produced at an appreciable rate; see Section 1.12) and, also, to desorb argon atoms embedded in the substrate itself. Argon etching needs to be repeated many times, since annealing may lead to segregation of impurities from the bulk of the sample to the surface. The driving force for segregation is the lowering in surface energy and, hence, is thermodynamic in origin. Periods of between several hours and several days of etching/anneal cycles may be required to obtain a spectroscopically pure surface (typically less than 1 atomic per cent of impurities).

Hence, surface science studies of gas–solid interfaces will generally involve the use of a vacuum apparatus consisting of a stainless steel chamber incorporating gauges to monitor the system pressure, argon ion etching facilities for surface cleaning, and a range of surface analysis probes. A variety of pumps and protocols are available for producing the UHV environment and the reader is referred to ref. 8 for more detailed information on these topics.

1.10 The problem of surface sensitivity

Although there are exceptions (for example, when one considers very small nanometre diameter particles), two factors mitigate against the study of gas–solid surfaces:

(i) the absolute number of atoms in a surface is small (problem of sensitivity); and

(ii) the ratio of surface atoms to bulk atoms is also very small (problem of selectivity).

To quantify each of these, an estimate will first be made of the absolute number of atoms in a surface. Second, the ratio of the number of surface to bulk atoms will be calculated.

(i) Imagine a cube of copper of side 1 cm^2 (copper lattice constant $a = 3.61$ Å). If this cube is oriented with a surface of (100) symmetry, the area of each primitive unit cell is given by

$$\left(\frac{3.61\ \text{Å}}{\sqrt{2}}\right)^2 = 6.52\ \text{Å}^2 = 6.52 \times 10^{-20}\ \text{m}^2$$

nearest neighbour
copper–copper distance $= a/\sqrt{2}$ (see Figure 1.14)

But, from Fig. 1.14, each primitive unit cell contains just one copper atom (although there are four atoms contributing to the cell, each of these atoms is shared by four unit cells and so contributes only $\frac{1}{4}$ of an atom to the unit cell in question). Hence, the total number of unit cells in 1 cm^2 of Cu(100) is

$$1\ \text{cm}^2/6.52 \times 10^{-16}\ \text{cm}^2 = 1.53 \times 10^{15}$$

$\therefore$ since there is one copper atom per unit cell, the total number of copper atoms in 1 cm^2 of Cu(100) is also 1.53×10^{15}. This represents $1.53 \times 10^{15}/6.02 \times 10^{23}$ mol cm^{-2} $= 2.5 \times 10^{-9}$ mol cm^{-2}. Detection of such small amounts of material is beyond the scope of most standard analytical methods. In addition, since the area of a surface probed can be much less than a full 1 cm^2 (often nearer to 1 mm^2) and sensitivity down to 1% of a monolayer is needed to define surface cleanliness, detection limits to picomolar (10^{-12} M) amounts are necessary!

(ii) In order to calculate the ratio of surface to bulk atoms in the copper cube, it is necessary to note that the interlayer spacing in the [100] direction [i.e. between (100) planes] is $a/2 = 3.61\text{Å}/2 = 1.805\ \text{Å} = 1.805 \times 10^{-8}$ cm. $\therefore$ Total number of (100) atomic layers in the copper cube is 1 cm/1.805×10^{-8} cm $= 5.54 \times 10^7$

Thus, it is seen that the surface:bulk ratio in a solid is typically $1{:}10^7$ to $1{:}10^8$, the exact value being dependent upon the surface structure of the element/compound under scrutiny. Therefore, if the surface component of any analytical signal is not to be swamped by the bulk signal, analytical methods (sensitive down to 10^{-12} M) that probe the top few atomic layers of a material need to be developed. Fortunately, spectroscopies based on the interaction of electrons with matter meet precisely the criteria outlined above and are wholly compatible with the UHV environment.

1.11 The interaction of electrons with matter

If a monoenergetic primary beam of electrons of kinetic energy, E_p, is incident upon a solid surface, it is found that a small percentage of those electrons are back-scattered from the surface without any loss of energy (elastically scattered). However, the majority of incident electrons, because they interact strongly with matter, lose their energy in a series of discrete (and sometimes continuous) 'energy loss' events, resulting in a broad spectrum of kinetic energies between zero and E_p. Figure 1.22 illustrates this phenomenon more clearly. Three energy loss mechanisms can be highlighted.

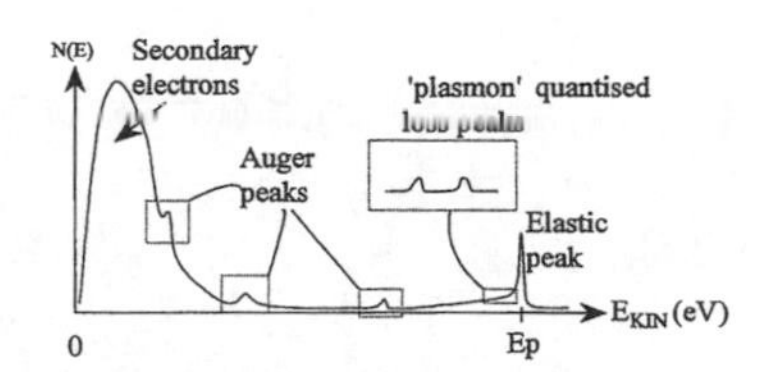

Fig. 1.22 Spectrum of number of electrons as a function of kinetic energy for an electron beam incident on a surface with initial energy E_p.

(i) **Plasmon excitation**. These are quantized electron density oscillations of the freely moving valence/conduction electrons of the solid. The

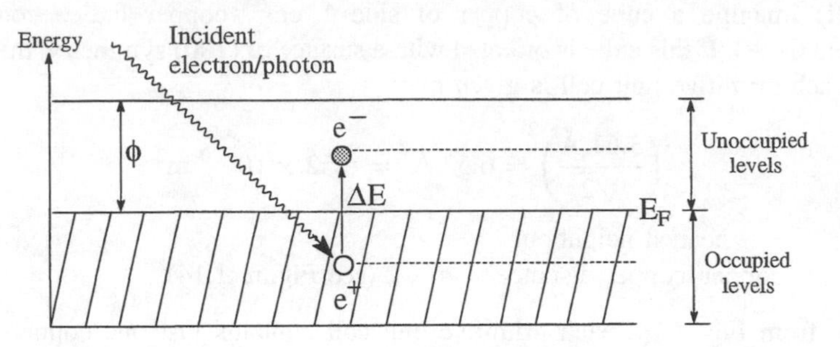

Fig. 1.23 Schematic diagram showing electron–hole pair excitation in a metal. ϕ = work function (see section 2.5)

quantum of energy involved in plasmon excitations is typically tens of eV depending upon the nature of the substrate.

(ii) **Electron-hole pair formation**. These excitations consist of promoting an electron from a filled to an empty electron state as illustrated in Fig. 1.23. Since valence levels in a solid form a continuous band of energy, a continuous energy loss range of 0–10 eV is possible.

(iii) **Vibrational excitation**. The vibrations of atoms in a solid are both quantized and coupled. Electron energy losses may be brought about by exciting these quantized vibrations of the solid lattice (phonon excitation) in addition to excitations of discrete vibrational modes on any adsorbate molecule. Phonon and vibrational excitations are relatively small in magnitude (0.5–0.01 eV).

Thus, by undergoing a combination of these various energy loss processes, electrons are removed from the incident beam as it passes through the solid. Electrons having undergone energy loss processes now appear at lower kinetic energies in the featureless background of the spectrum depicted in Fig. 1.22 (secondary electrons). Hence, the intensity of electrons of well-defined kinetic energy will be damped as a function of distance into the solid. The intensity decay follows an exponential first-order decay law typical for the travel of radiation through matter (*cf.* Beer law for the absorbance of electromagnetic radiation)

$$I(d) = I_0 \exp\left(\frac{-d}{\lambda(E)}\right) \tag{1.53}$$

where $I(d)$ is the intensity after the primary electron beam has travelled a distance, d, through the solid and I_0 is the initial beam intensity before interaction with the solid. The decay of intensity is critically dependent upon the parameter, $\lambda(E)$, termed the **inelastic mean free path** (IMFP). This is defined as the distance an electron beam can travel before its intensity decays to 1/e or (1/2.718) of its initial value. The inelastic mean free path is an index of how far an electron can travel on average before losing energy. A small value of λ indicates a high probability for energy loss and the ability to travel only a short distance before being absorbed. Thus, electrons with a small inelastic mean free path are highly surface sensitive.

Suppose we wish to determine the distance into the solid necessary to quench the intensity of the incident electron beam to 90% of its original value.

Assuming an IMFP of 5 Å, $I_0 = 100\%$, $I(d) = 10\%$ (remember, that if 90% of intensity is lost then 10% remains!), then from eqn 1.53, $d = 11.5$Å; d is of the order of a few interatomic spacings, hence the majority of electrons in the incident electron beam have been absorbed within the selvedge region. The corollary of this statement is that any electrons generated in the solid with an IMFP of 5 Å that manage to escape into the vacuum must have originated from the first few surface atomic layers.

Hence, to discuss quantitatively the surface sensitivity of electrons, we must know the value of the inelastic mean free path of an electron in the material of interest. It is generally accepted that the inelastic mean free path is only weakly material dependent. However, it is strongly dependent upon the electron kinetic energy. Figure 1.24 illustrates the result of a range of measurements of $\lambda(E)$ for metals. The solid line describes a so-called 'universal curve' for metals. Other 'universal curves' have been constructed for other types of material, such as hydrocarbons and semiconductors. An empirical relationship fitting the data in Fig. 1.24 has been suggested by Seah and Dench [9]:

$$\frac{\lambda}{\text{nm}} = \frac{538\text{a}}{E^2} + 0.41\text{a}^{3/2}\,(E_{\text{kin}})^{1/2} \qquad (1.54)$$

where E is the kinetic energy of the electrons, in electron volts, and a* is the mean atomic diameter of the element in nanometres.

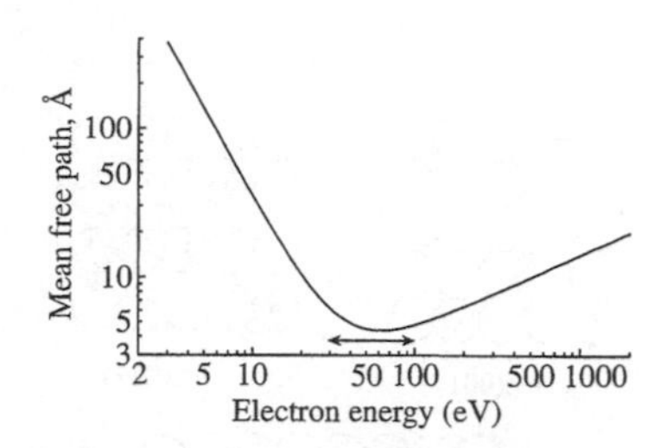

Fig. 1.24 'Universal curve' showing, schematically how the inelastic mean free path of an electron in a solid varies with its kinetic energy. The double arrow indicates the range where surface sensitivity is greatest. Note that accurate values of the inelastic mean free path may be obtained using equation 1.54.

The shape of the curve may be rationalized as follows: for kinetic energies greater than approximately 50 eV the first term in eqn 1.54 may effectively be ignored and the mean free path is proportional to the square root of the kinetic energy and, consequently directly proportional to the electron velocity. This is understandable since the faster an electron travels, the shorter the time taken in passing through a given thickness of solid and, hence, the less likely is the chance of energy loss. The inelastic mean free path passes through a minimum at about 50 eV and rapidly increases at kinetic energies below about 30 eV. This is due to the removal of the dominant energy loss mechanism—that of plasmon excitation. As the kinetic energy falls below the critical energy necessary for plasmon excitation, the mean free path increases dramatically. The first term in eqn 1.54 models this rapid increase.

*$\text{a} = (\Omega/1000\rho L)^{1/3}$
where Ω = molar mass (g mol^{-1}),
ρ = density (kg m^{-3})
L = Avogadro number (mol^{-1})
a = mean atomic diameter (m).

A simple yet widely utilized method to enhance surface sensitivity may also be mentioned. While the discussion to date has assumed that the electron beam is incident normal to the analysed surface, variation of the angle of incidence and/or emission will, by definition, increase the effective path length in the solid as indicated in Fig. 1.25. For angles of incidence other than normal to the surface

$$I(d) = I_0 \exp\left(\frac{-d}{\lambda \cos\theta}\right) \qquad (1.55)$$

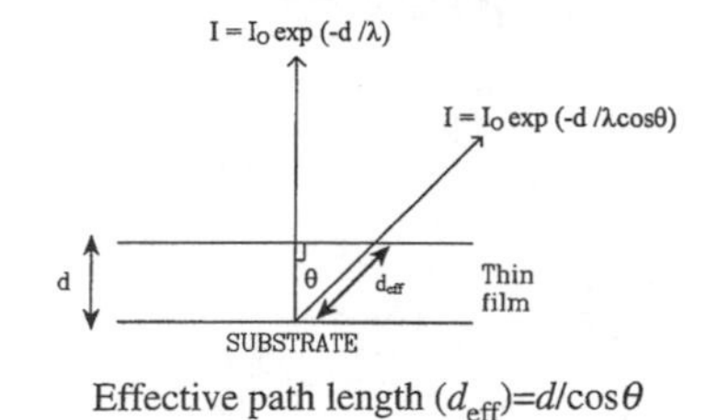

Fig. 1.25 Figure showing how the effective path length (d_{eff}) for an electron in a solid will increase on going from normal to glancing angles of incidence/emission.

Clearly, the larger the angle of incidence/emission, the larger the effective path length and the more likely is the process of energy loss. Hence, a grazing incidence geometry will increase surface sensitivity in electron-based spectroscopies (see also Fig. 2.6).

1.12 Mobility in two dimensions

In order for surface systems to attain their equilibrium configuration, it is necessary for surface atoms and molecules to be sufficiently mobile to reach their minimum free energy state. The process of migration of atoms/molecules across a surface is termed 'surface diffusion'. For a given substrate/adsorbate combination, the rate of diffusion is critically dependent upon

(a) the crystallographic direction along which diffusion occurs;

(b) the absolute temperature; and

(c) the surface coverage of adsorbate.

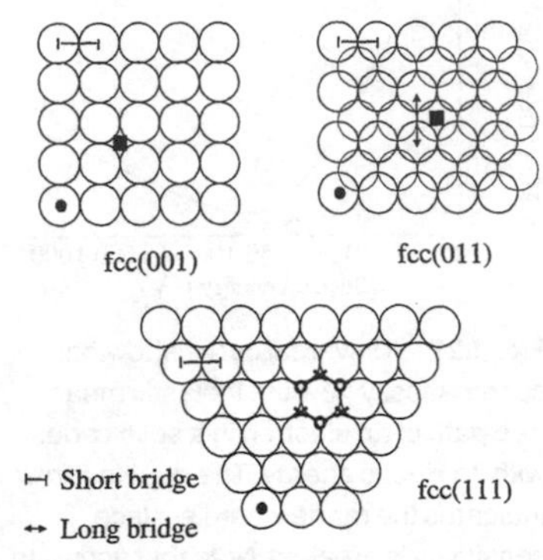

Fig. 1.26 The various sites available for adsorption on fcc surfaces. Note that two types of three-fold hollow may be distinguished on fcc (111) surfaces.

Since the arrangement of adsorbate molecules on a surface is a key parameter governing surface reactivity, a brief exposition of the salient features of surface diffusion will be given. Figure 1.26 shows the principal adsorption sites on fcc low index surfaces and Fig. 1.27(a) shows a top view of the outermost layer of a fcc (110) surface.

Consider an adsorbate whose equilibrium position is a hollow site. In order to diffuse to a neighbouring site as illustrated by the dotted circles in Fig. 1.27(a), the adatom must pass over the bridge sites, which possess a lower binding energy and hence a higher potential energy. Figure 1.27(b) illustrates schematically the potential energy profile along the dashed line, i.e. along the $[1\bar{1}0]$ channel. Because the adsorbed particle has to pass over an energy barrier in order to migrate to a neighbouring site, it is an activated process and generally follows an Arrhenius-like behaviour

$$D(\theta, T) = D_0 \exp\left(\frac{-E_{\mathrm{ACT}}(\theta)}{RT}\right) \tag{1.56}$$

where $D(\theta, \mathrm{T})$ is the diffusion coefficient ($\mathrm{cm^2\ s^{-1}}$); E_{ACT} is the activation energy barrier ($\mathrm{kJ\ mol^{-1}}$); T is the absolute temperature (K); and D_0 is a pre-exponential factor called the 'diffusivity' ($\mathrm{cm^2\ s^{-1}}$).

The diffusivity is related to the entropy change between the equilibrium hollow site and the 'activated complex'. An approximation often assumed is that the entropies in the equilibrium site and the activated complex are the same. In such cases, D_0 may be assumed to be $\simeq 10^{-2}\ \mathrm{cm^2\ s^{-1}}$. The diffusion coefficient allows calculation of the average distance travelled, $\langle x \rangle$, in a time t, assuming diffusion occurs *via* a 'random walk' process, i.e. a process in which the probability of hopping in opposite directions along the channel is equal

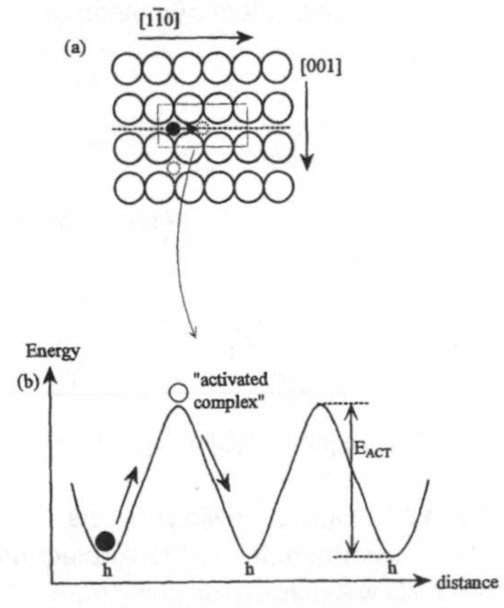

Fig. 1.27 (a) Diffusion of an adatom between adjacent fourfold hollow sites along the $[0\bar{1}1]$ direction. (b) The corresponding potential energy diagram for diffusion in this direction. h = fourfold hollow site.

$$\langle x \rangle = (Dt)^{1/2} \tag{1.57a}$$

$$\text{and} \quad D = \langle x \rangle^2 / t \tag{1.57b}$$

Returning to the example of diffusion on a fcc (110) surface, one can see that, although the equilibrium site is the same for diffusion in the orthogonal [001] direction (at 90° to $[1\bar{1}0]$), the 'activated complex' is an adatom bonded either at a short bridge or atop site. As the aforementioned sites will have a different energy to the long bridge site, the activation barrier and hence the diffusion coefficient, will be different in the two crystallographic directions, and diffusion must therefore be an anisotropic process. Diffusion coefficients are determined by measuring the rate of spreading of an adsorbate across a surface for a range of temperatures.

Taking natural logs of eqn 1.56 and substituting 1.57b, we obtain

$$\ln\left(\frac{\langle x\rangle^2}{t}\right) = \ln D_0 - \frac{E_{ACT}}{R}\left(\frac{1}{T}\right) \tag{1.57c}$$

Thus, a plot of $\ln \frac{\langle x\rangle^2}{t}$ *versus* $1/T$ should yield a straight line if diffusion is Arrhenius-like, with a gradient equal to $-E_{ACT}/R$, and an intercept on the y-axis of $\ln D_0$, as shown in Fig. 1.28.

Table 1.5 lists a range of activation energies for surface diffusion for different adsorbate–substrate combinations.

Table 1.5 Activation barrier for diffusion and the corresponding value of the diffusivity [2]

System	E_{ACT} (kJ mol^{-1})	D_0 (cm^2 s^{-1})
O/W {110}	59	1×10^{-7}
Xe/W {110}	5	7×10^{-8}
H/Ni{100}	15	2×10^{-3}
CO/Ni {100}	20	5×10^{-2}
Ni/Ni {100}	159	300

Physisorbed molecules with low heats of adsorption have low diffusion activation energies and, hence, are highly mobile at ambient temperatures, making thousands of atomic hops per second, while more strongly bound chemisorbed species, with higher activation barriers, remain virtually immobile at similar temperatures. Thus, whether or not surface equilibrium is obtained at a given temperature is critically dependent upon the particular adsorbate/substrate combination. In addition, it is found that the activation energy for diffusion on average is larger on atomically rough surfaces than on smoother, close-packed surfaces. For example, for fcc surfaces, the rate of diffusion as measured by the diffusion coefficient increases in the order:

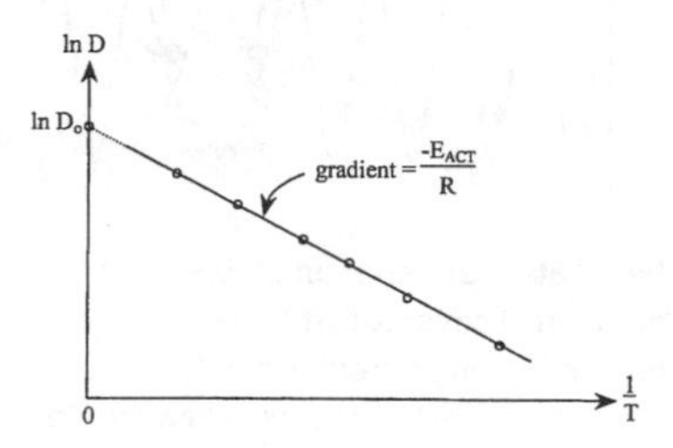

Fig. 1.28 Determination of E_{ACT} and D_0 from measurements of the diffusion coefficient as a function of temperature.

stepped < (100) < (110) < (111)

$\longrightarrow$ increasing diffusion coefficient

$\longleftarrow$ increasing surface 'roughness'

It is relatively easy to rationalize the high activation energy across a surface step. Figure 1.29 describes a one-dimensional potential energy profile in the direction of a surface step. The increased coordination offered by the step site means that an adatom is generally bound more strongly at a step site than on the atomically flat terrace, often leading to nucleation of adsorbed overlayers at these 'high surface energy sites'. A corollary of this is that the activation energy for diffusion from a step to a terrace site is considerably larger than that between adjacent terrace sites, leading to a lowering of the diffusion constant and hence a reduced mass transport in a direction orthogonal to the step direction. Obviously, the more atomic steps per unit length, i.e. the higher the step density, the slower the net rate of diffusion.

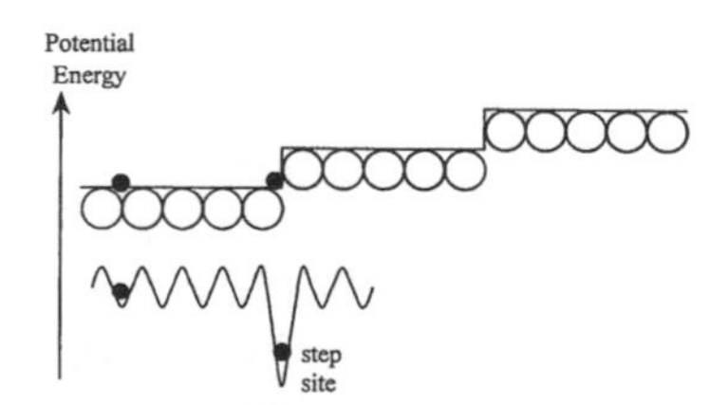

Fig. 1.29 One-dimensional energy profile in the direction of a surface step. Note, the larger activation energy barrier for diffusion from step to terrace sites as compared to diffusion on terraces.

In the limit of vanishing adsorbate coverage, experimental determination of the diffusion activation energy in a well-defined crystallographic direction yields information on the 'potential energy hypersurface', i.e. the energy difference between adsorbates bonded at different positions in the substrate two-dimensional unit cell. This is because diffusion activation energies are a measure of the binding energy *difference* between adsorption sites.

Using the data in Table 1.5, calculate the average distance moved by a physisorbed xenon atom on W(110) at 100 K. How does this compare with that for an oxygen atom at the same temperature?

However, the situation becomes more complex at finite coverage, when the statistical probability of an adsorbate having other adsorbates in its local vicinity becomes high (interatomic separations ≤ 10 Å). Experimental studies have indicated that the activation energy for diffusion can be strongly dependent upon surface coverage. Figure 1.30(a) shows schematically the activation energy for diffusion of atomic oxygen on a tungsten body-centred cubic (100) surface as a function of oxygen coverage. For surface coverages up to 0.5 ML, the activation energy for diffusion remains relatively constant.

The symbol ML is often used instead of 'monolayers'.

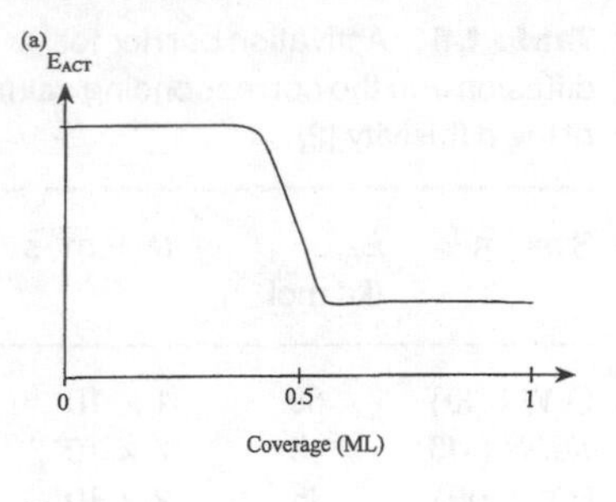

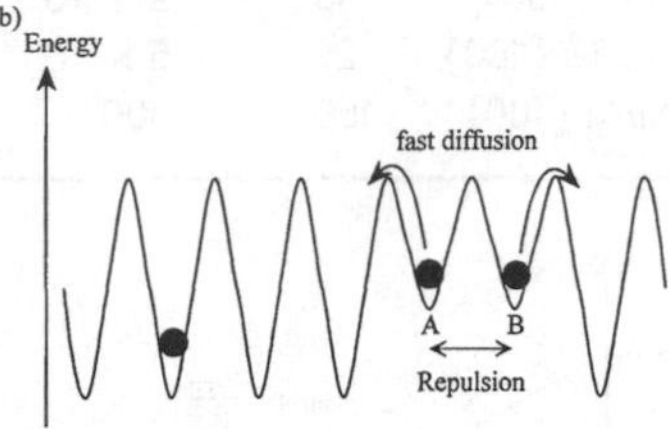

Fig. 1.30 (a) Variation of E_{ACT} as a function of coverage. (b) The corresponding potential energy diagram. Note that at coverages greater than 0.5 ML, nearest–neighbour sites become occupied leading to a decrease in E_{ACT} owing to repulsive lateral interactions.

This is because, as long as the oxygen coverage remains at or below 0.5 ML, no oxygen atoms need to occupy nearest–neighbour sites. Figure 1.30(b) illustrates the corresponding potential energy profile for this surface. Note that, on the left of the diagram, atoms do not occupy nearest–neighbour sites. However, on the right-hand side, two oxygen adatoms (A and B) are shown having been forced into nearest–neighbour sites. The strong mutual repulsion between adsorbates A and B leads to an increase in the potential energy of these adsorbates relative to oxygen adatoms not occupying nearest–neighbour sites. Hence the rapid lowering of the activation energy for diffusion at $\theta \geq 0.5$ ML. Since surface diffusion is so important in many areas of surface reactivity, it is necessary to discuss the various interactions between adatoms and their rôle in modifying the rate of surface diffusion. Two aspects of the above will be discussed. First, the types of intermolecular forces exhibited by adsorbed molecules and, secondly, the net attractive/repulsive interactions that result.

1.13 Lateral interaction (forces between adsorbate molecules)

There are four types of adsorbate–adsorbate interactions to be considered

- direct Coulombic interactions
- van der Waals forces
- covalent/metallic bonding interactions
- indirect substrate-mediated forces

Direct Coulombic interactions

These occur for adatoms that have undergone charge transfer with the substrate or, alternatively, include strongly electropositive or electronegative atoms within a molecular adsorbate. Adsorption of alkali metals is a good example of adsorbates forming a strongly polar bond with the substrate at low coverages. As illustrated in Fig. 1.31, charge transfer from the alkali metal to the substrate occurs, leaving the adsorbate with a partial positive charge. The force of repulsion may be estimated by application of Coulomb's law

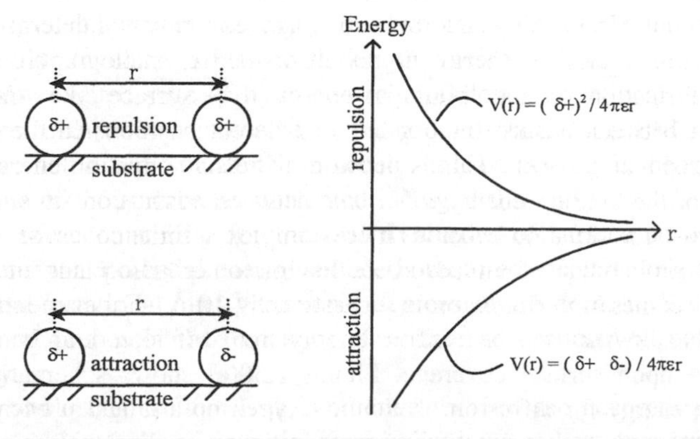

Fig. 1.31 Variation in electrostatic interactions as a function of adsorbate separation for adsorbates forming strongly polar bonds to the surface.

$$F = \frac{|\delta^+|^2}{4\pi\varepsilon r^2} \tag{1.58}$$

Hence, the potential energy of interaction is given by

$$\text{Energy} = -\int_{r=\infty}^{r} F\mathrm{d}r = \frac{-1(\delta^+)^2}{4\pi\varepsilon}\int_{\infty}^{r}\frac{1}{r^2}\mathrm{d}r = \frac{|\delta^+|^2}{4\pi\varepsilon r} \tag{1.59}$$

Thus, according to simple electrostatics, the interaction energy will be proportional to $1/r$. Also, since the energy is positive, the nature of the interaction is repulsive. It is difficult to calculate the exact interaction energy for a number of reasons. First, the substrate electrons tend to respond by 'shielding' (reducing) the electrostatic field set up by the adsorbate. In addition, the exact degree of polarity in the surface–adsorbate bond, i.e. the magnitude of the charge residing on the adatom, is difficult to quantify. Nevertheless, the general argument holds that repulsive interactions will lead to a situation where the adsorbates maximize their average separation by forming a two-dimensionally dispersed phase in order to *decrease* the interaction energy (r large in eqn 1.59).

In certain cases, Coulombic interactions may be attractive, resulting in the formation of densely packed two-dimensional structures. This is often the case when two or more adsorbates are co-adsorbed and when the individual co-adsorbates themselves have an opposing direction of charge transfer with the substrate. Figure 1.32 illustrates an example involving the co-adsorption of a strongly electropositive element (potassium) with a strongly electronegative element (oxygen). Such systems tend to form intimately mixed structures in which the ionic interactions between oppositely charged adsorbates are maximized. In this case, the electropositive potassium atoms are surrounded by four electronegative oxygen atoms in a surface analogue of a two-dimensional ionic salt.

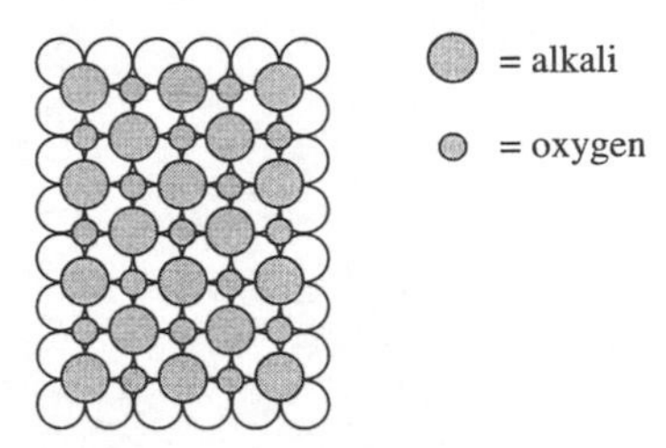

Fig. 1.32 Co-adsorption of electropositive and electronegative adsorbates to generate an intermixed phase. This phase is a surface analogue of a two-dimensional salt.

Direct covalent/metallic bonding

This can occur between two adsorbates as long as each possesses a partially filled valence orbital. This type of bonding is prevalent for adsorbates exhibiting a similar electronegativity to the substrate, such that there is little polar character in the substrate–adsorbate bond, and also for adsorbates that remain 'unsaturated' even when adsorbed on a surface. A classic example of such behaviour is the adsorption of a transition metal adatom on to a metal surface. Strong attractive lateral interactions between adatoms arise as a result of the possible energy minimization upon metal–metal bond formation within the adsorbate layer. Metals adopt high coordination structures both in bulk phases and on surfaces in which a large number of bonds are formed between nearest neighbours, leading to the formation of a 'giant' metallic network. For a metal overlayer, this results in the formation of a two-dimensional adsorbate band structure with significant delocalization of adsorbate valence electrons.

Van der Waals forces

These are generally attractive in origin. Such interactions occur in so-called self-organizing or self-assembling monolayers. Van der Waals forces originate from instantaneous distortions in the electron clouds of atoms and molecules,

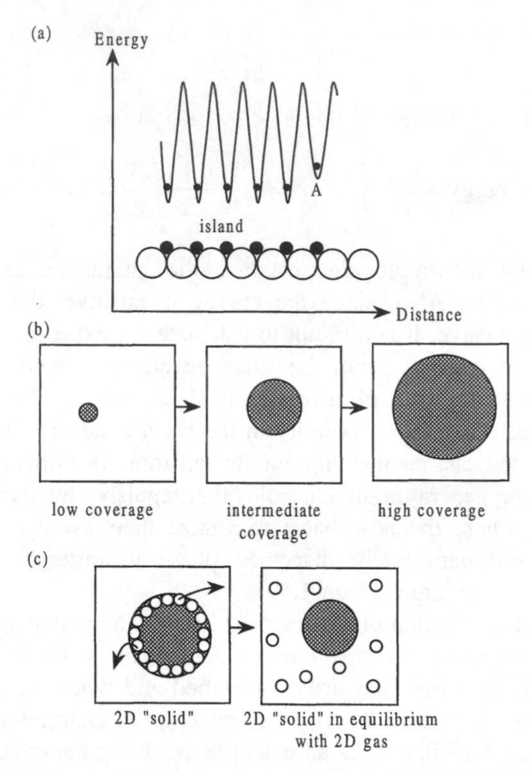

Fig. 1.33 (a) Potential energy diagram for atoms exhibiting attractive lateral interactions and forming a two-dimensional island. Note that edge atoms possess a higher potential energy than those within the island. (b) At 0 K, the island size increases with coverage (Ostwald ripening). (c) At finite temperatures, edge atoms may detach themselves from the island forming a two-dimensional gas in equilibrium with a two-dimensional solid (the island size decreases).

which subsequently induce a temporary dipole moment within a neighbouring molecule. Molecules with a high molecular weight (possessing a large number of loosely bound valence electrons) exhibit large van der Waals attractions. In general, these interactions are considerably weaker than either of the previous two cases and, hence, are only dominant in adlayers of large non-polar adsorbates.

Indirect interactions

These are also possible when the interaction is mediated by the substrate electrons. However, these are complex in nature and beyond the scope of an introductory text.

1.14 The effect of lateral interactions on the distribution of adsorbates

Net attractive lateral interactions

Attractive lateral interactions lead to a distribution of adsorbed species in the form of 'two-dimensional islands' exhibiting 'local' coverages considerably higher than the overall coverage averaged over the entire surface. Figure 1.33(a) illustrates a simplified one-dimensional energy profile in which six adsorbed atoms are arranged into a small 'island'. The adatom labelled A has a higher potential energy compared with atoms within the island. This follows because the lowering in potential energy (PE) that accompanies island formation arising from attractive interactions is not available to an isolated adatom (e.g. through direct metallic/covalent bonding).

When adatom A joins the island, its potential energy will decrease. However, since it lacks a nearest neighbour, its energy is still higher than atoms in the interior of the island which possess two nearest neighbours. Its stability will only be maximized when an additional atom joins the island and fills the empty nearest–neighbour site, i.e. when it is no longer the 'edge' atom. Thus, for systems with net attractive interaction on a two-dimensional perfect, defect–free surface in thermal equilibrium at 0 K, the lowest energy state is for adsorbates to condense into one large, circular two-dimensional island. In this way the number of high energy adsorbates at the island edge is minimized. Formation of large islands from small ones is known as 'Ostwald ripening'. As the surface coverage is increased, the island will simply increase in diameter, as shown in Fig. 1.33(b).

Even if one could produce such a two-dimensionally perfect surface, it is only at very low temperatures (where entropy plays a small role) that the distribution shown in Fig. 1.33 would occur. It is the Gibbs free energy (G) of the surface system that must be minimized to yield the equilibrium configuration

$$G = H - TS \tag{1.60}$$

While the enthalpy term (H) is minimized for packing in a single, circular island, such a distribution has a low entropy (S) since it represents the most ordered surface phase. Thus, while at absolute zero, minimization of the enthalpy term is the only criterion that must be considered, at finite temperatures the $T\Delta S$ term favours break-up of the island, since $\Delta S = S_{\text{random}} - S_{\text{island}}$ is positive. As the temperature increases, there will be an increasing driving force for adatoms to detach themselves from the island and form a 'two-dimensional gas', as shown in Fig. 1.33(c). This is an example of a two-dimensional surface phase transformation.

Real surfaces are far from ideal and contain atomic steps and a range of other defects. Hence, instead of one large, two-dimensional island, a surface adsorbate generally consists of an array of islands of diameters between 20 and several hundred Ångstroms, separated by areas of clean surface. As the surface coverage is increased, the island density rises, until the whole surface is covered with an array of densely packed islands with boundaries between them.

The arrangement of adsorbed atoms and molecules within the island relative to the underlying substrate (the superlattice) may be termed: (i) commensurate; or (ii) incommensurate.

A **commensurate** overlayer forms when the substrate–adsorbate interaction tends to dominate over any lateral adsorbate–adsorbate interaction. Each adsorbate adopts an interadsorbate separation that is either equal to that of the substrate adatoms or a simple multiple of the substrate spacing. In the case of an **incommensurate** overlayer, the adsorbate–adsorbate interactions are of similar magnitude to those between adsorbate and substrate, and the spacing adopted is a compromise between maximizing both adsorbate–substrate and

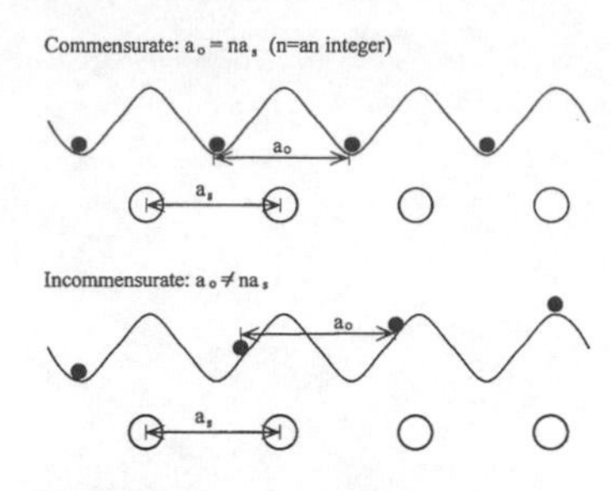

Fig. 1.34 Commensurate and incommensurate surface phases. Open circles represent substrate atoms. Filled circles correspond to adsorbate molecules. a_s = interatomic spacing of substrate surface atoms. a_0 = interatomic spacing of overlayer atoms.

adsorbate–adsorbate interactions. This leads to an adsorbate–adsorbate spacing that may not be a simple multiple of the substrate lattice spacing and a wide range of different local adsorption sites will be populated. This is illustrated in Fig. 1.34.

Repulsive lateral adatom–adatom interactions

Adsorbates exhibiting net repulsive lateral interactions tend to form dispersed phases since, for a given surface coverage, the system's free energy may be minimized by maximizing the average distance between adsorbates (*cf* alkali metal adsorption (Sec. 1.13)). Such systems often form ordered 'superstructures' which, at least for low and medium surface coverages, are usually commensurate with the underlying lattice. However, as adsorbate–adsorbate interactions increase (high coverage), incommensurate phases may form.

Figure 1.35 illustrates adsorption on to a square substrate of an adsorbate exhibiting repulsive interactions. The formation of a number of commensurate superstructures is evident in which the average distance between neighbouring adatoms decreases with increasing surface coverage (1/9 monolayer → 1/2 monolayer). In addition, the **matrix notation** describing the overlayer is given.

1.15 Naming overlayer structures

There are two ways of naming overlayer structures. The first of these, the **Wood's notation** is suited particularly to commensurate structures and relates the overlayer **mesh** (or **net**) to the substrate mesh. For example, if one takes the phase corresponding to $\theta = 1/2$ in Fig. 1.35, which depicts a simple overlayer on a square substrate, it is seen that the unit cell describing the overlayer is centred with unit vectors ($\boldsymbol{a}_0$ and $\boldsymbol{b}_0$) parallel to $\boldsymbol{a}_s$ and $\boldsymbol{b}_s$ respectively, the vectors defining the primitive unit cell of the substrate.

In addition

$|x|$ = magnitude of vector x

$$|\boldsymbol{a}_0| = 2|\boldsymbol{a}_s| \quad \text{and} \quad |\boldsymbol{b}_0| = 2|\boldsymbol{b}_s|$$

That is, $\boldsymbol{a}_0$ is twice as long as $\boldsymbol{a}_s$, and $\boldsymbol{b}_0$ is twice as long as $\boldsymbol{b}_s$. Hence, the overlayer unit cell is referred to as a (2 × 2) (two by two) overlayer. However, because the overlayer mesh contains an atom in the centre, its full description would be 'c(2 × 2)' (centred two by two). We could equally well define the unit cell of the overlayer in terms of a primitive cell (vectors $\boldsymbol{a}'_0$ and $\boldsymbol{b}'_0$). Note that $\boldsymbol{a}'_0$ and $\boldsymbol{b}'_0$ are now rotated with respect to the substrate by 45°. Also, $|\boldsymbol{a}'_0| = \sqrt{2}|\boldsymbol{a}_s|$ and $|\boldsymbol{b}'_0| = \sqrt{2}|\boldsymbol{b}_s|$. Hence, an alternative name for this overlayer would be a $(\sqrt{2} \times \sqrt{2})$R45° (root two by root two rotated 45°) structure. To summarize, the Wood's notation may be described fully as

$$\mathrm{M}(hkl)\left(\frac{|\boldsymbol{a}_0|}{|\boldsymbol{a}_s|} \times \frac{|\boldsymbol{b}_0|}{|\boldsymbol{b}_s|}\right) - \mathrm{R}\alpha^\circ - \mathrm{A}$$

where M = chemical symbol of substrate; $(h\ k\ l)$ = Miller index of surface plane; $|\boldsymbol{a}_s|$ and $|\boldsymbol{b}_s|$ = magnitude of substrate net vectors; $|\boldsymbol{a}_0|$ and $|\boldsymbol{b}_0|$ = magnitude of surface overlayer net vectors; α = angle between substrate and overlayer meshes (omitted if $\alpha = 0°$); and A = chemical symbol of surface species (omitted if A = M, clean surface).

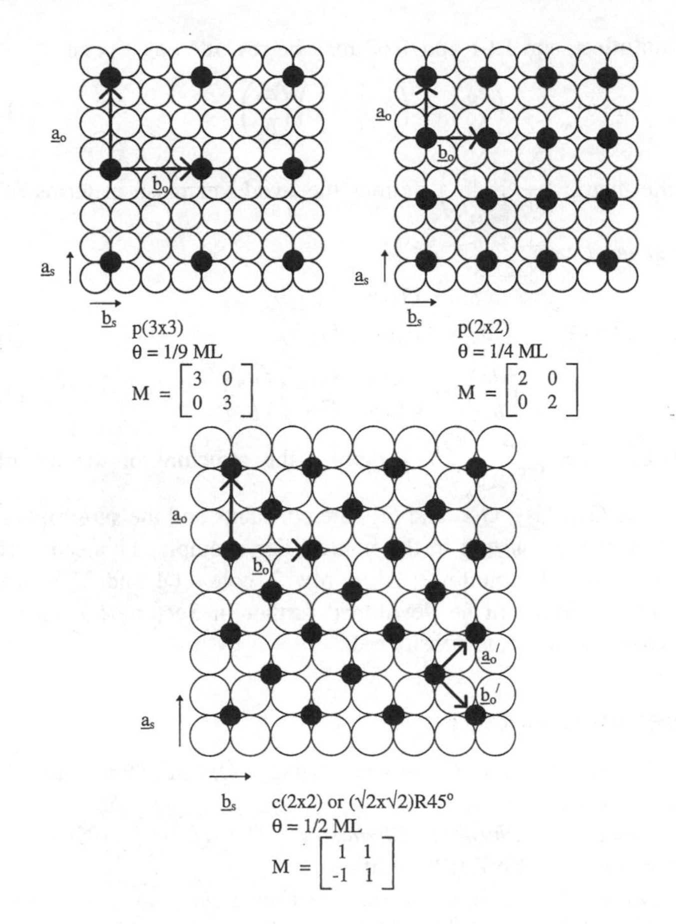

Fig. 1.35 Changes in the surface geometry as a function of coverage for an adsorbate exhibiting repulsive lateral interactions.

Hence, the high coverage structure depicted in Fig. 1.35 may equally well be referred to as either a c(2 × 2) or $(\sqrt{2} \times \sqrt{2})$R45° [strictly p$(\sqrt{2} \times \sqrt{2})$ R45° with 'p' for primitive—but the absence of the p is usually taken to mean a primitive cell] in the Wood's notation. It is simply a question of whether or not one chooses a primitive unit cell to describe the overlayer mesh. The c(2 × 2) structure depicted in Fig. 1.35 is a commonly observed phase for adsorbates on square lattices and is seen, for example, when sulfur is adsorbed on Pt(100) up to a coverage of 0.5 monolayers. In the Wood's notation, the sulfur phase would be described as Pt(100)$(\sqrt{2} \times \sqrt{2})$R45° – S – (0.5ML).

A more general notation, known as 'matrix' notation may be used to describe overlayer structures that are both commensurate *and* incommensurate. The first step is to define the primitive overlayer mesh vectors ($\boldsymbol{a}_0$ and $\boldsymbol{b}_0$) in terms of a linear combination of the **primitive** unit mesh vectors of the substrate ($\boldsymbol{a}_s$ and $\boldsymbol{b}_s$). For example, if one takes the $(\sqrt{2} \times \sqrt{2})$R45° phase in Fig. 1.35 once again

$$\boldsymbol{a}'_0 = 1\boldsymbol{a}_s + 1\boldsymbol{b}_s \qquad (1.61)$$

and

$$\boldsymbol{b}'_0 = -1\boldsymbol{a}_s + 1\boldsymbol{b}_s \qquad (1.62)$$

In **matrix notation**, eqn 1.61 and 1.62 may be rewritten as

$$\begin{pmatrix} a_0 \\ b_0 \end{pmatrix} = \begin{pmatrix} 1 & 1 \\ -1 & 1 \end{pmatrix} \begin{pmatrix} a_s \\ b_s \end{pmatrix} \qquad (1.63)$$

Note that the matrix $\begin{pmatrix} 1 & 1 \\ -1 & 1 \end{pmatrix}$ defines the overlayer mesh in terms of the substrate.

For the *general* case

$$a_0 = G_{11}a_s + G_{12}b_s$$

and

$$b_0 = G_{21}a_s + G_{22}b_s \qquad (1.64)$$

Hence

$$\begin{pmatrix} a_0 \\ b_0 \end{pmatrix} = \begin{pmatrix} G_{11} & G_{12} \\ G_{21} & G_{22} \end{pmatrix} \begin{pmatrix} a_s \\ b_s \end{pmatrix} \qquad (1.65)$$

and the matrix $G = \begin{pmatrix} G_{11} & G_{12} \\ G_{21} & G_{22} \end{pmatrix}$ defines the structure of the adsorbate overlayer where G_{11}, G_{12}, G_{21}, and G_{22} are constants and the subscripts refer to the position of the constant in the matrix. For example, 11 means row 1, column 1; 12 = row 1, column 2; 21 = row 2, column 1 and 22 = row 2, column 2. These ideas will be developed further in Section 2.3 when low energy electron diffraction is considered.

References for Chapter 1

1. P.W. Atkins, *Physical Chemistry*, (1982), Oxford University Press, Oxford, (second edition).
2. A.W. Adamson, *Physical Chemistry of Surfaces*, (1990), Wiley Interscience, New York (fifth edition).
3. R. Aveyard and D.A. Haydon, *Introduction to the principles of surface chemistry*, (1973), Cambridge University Press, Cambridge.
4. H. Conrad, G. Ertl, J. Koch and E. Latta, *Surf. Sci.*, **43** (1974) 462.
5. M.A. Morris, M. Bowker and D.A. King in *Comprehensive chemical kinetics*, eds. C.H. Bamford, C.F.H. Tipper and R.G. Compton, (1984), Elsevier, Amsterdam.
6. R.A. Van Santen and J.W. Niemantsverdriet, in *Fundamental and applied catalysis*, eds. M.V. Twigg and M.S. Spencer, (1995), Plenum, New York.
7. E. Preuss, B. Kraul-Urban and R. Butz, *Kernforschunganlage Julich Laue Atlas*, (1973), Wiley, New York.
8. G.F. Weston, *Ultrahigh Vacuum Practice*, (1985), Butterworths, London.
9. *Practical surface analysis by Auger and X-ray photoelectron spectroscopy*, eds D. Briggs and M.P. Seah, (1993), Wiley, New York.

2 Spectroscopic techniques for probing solid surfaces

While classical thermodynamic methods, as discussed in Chapter 1, can provide useful surface information, such data tends to yield 'average' properties of the system rather than atomic/molecular specificity. A truly microscopic understanding of surface phenomena requires one to probe surface properties experimentally at the molecular level. In this chapter, a range of surface-sensitive spectroscopies, capable of providing detailed information concerning the geometric structure, chemical composition, and electronic properties of a solid surface, will be described. Probing a surface spectroscopically requires a stimulus leading to a measured response. The various stimulus/response combinations used in surface science studies are summarized in Fig. 2.1 (the Propst diagram). In principle, any combination may form the basis of a spectroscopy, e.g. **photoelectron spectroscopy** (see Section 2.1) would correspond to irradiation with photons (stimulus) leading to electron emission (response).

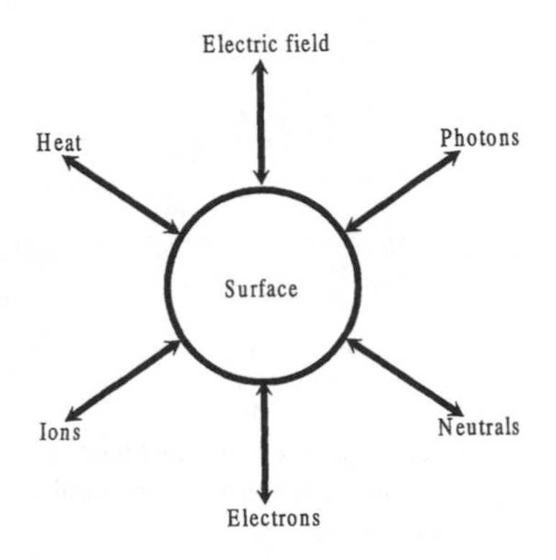

Fig. 2.1 The Propst diagram.

In order to characterize fully a surface we would like to answer a range of questions.

(i) What types of atoms are present at a surface and what is their surface concentration?

(ii) Where, precisely, are the atoms/molecules located on a surface and what bond lengths and bond angles do the molecules exhibit?

(iii) How strong is the bonding of adsorbate atoms to a surface and how does the nature of the surface bond influence surface reactivity?

No single spectroscopy is capable of answering adequately all of these questions, and modern surface analysis increasingly relies on a multi-technique approach in which a range of surface-sensitive probes are used in tandem to provide complementary information. In a text of this size it is impossible to discuss all of the surface spectroscopies available. However, there does exist a small nucleus of techniques which are viewed by many as the work horses of the field. Although selective, by narrowing attention to just a small number of experimental surface probes, a basic discussion of the physicochemical principles involved and the type of information they provide is possible. This chapter will begin by examining the two major techniques (X-ray photoelectron spectroscopy and Auger electron spectroscopy) utilized to gain information on surface chemical composition followed by a discussion of methods suitable for elucidating surface structure including, both a diffraction-based approach (low-energy electron diffraction) and more direct probes, based on electron tunnelling (scanning probe microscopies). Finally, spectroscopies to investigate surface chemical bond formation and reactivity, such as ultra violet photoemission, vibrational spectroscopies (reflection-absorption infra red and high resolution electron energy loss) the strength of surface bonding (thermal desorption spectroscopy), and surface reaction rates (molecular beam spectroscopies) will be considered.

2.1 X-ray photoelectron spectroscopy (XPS)

XPS is one of the most versatile techniques used for analysing surfaces chemically. The basis of the technique lies in Einstein's explanation of the photoelectric effect, whereby photons can induce electron emission from a solid provided the photon energy ($h\nu$) is greater than the work function (the work function of a solid is defined as the minimum energy required to remove an electron from the highest occupied energy level in the solid to the 'vacuum level', and is usually given the symbol ϕ; see Section 2.5). The vacuum level is the energy of an electron at rest (zero kinetic energy) in a vacuum far removed from neighbouring particles such that it has no interaction with them (zero potential energy). The vacuum level may be used as an 'energy zero'.

In XPS, a monochromatic beam of X-rays is incident upon a solid surface, causing photoemission from both core and valence levels of surface atoms into the vacuum. Core levels are defined as the inner quantum shells, which do not participate in chemical bonding, while valence levels are electrons in the more weakly bound, partially filled outer quantum shells. Figure 2.2 illustrates schematically the energetics of a photoemission experiment. The key to chemical identification is that core electrons deep inside atoms are largely insensitive to their surroundings when condensed into the solid phase and retain binding energies E_B that are signatures of the atom type, i.e. the number of protons in the nucleus. The outermost electrons, which participate in chemical bonding in a solid, are broadened into a 'valence band'. Emission from the valence band is most effectively probed by ultra violet photoemission

Electron energy levels may be labelled both in X-ray and spectroscopic notation as shown in Table 2.1.

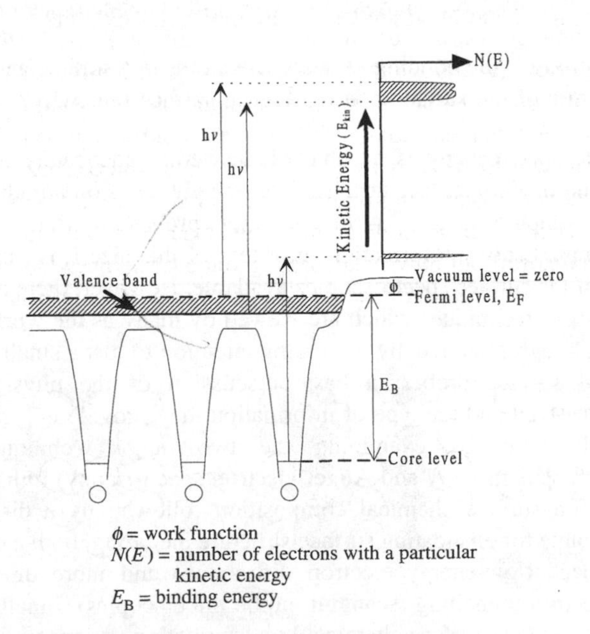

Fig. 2.2 The energetics of an X-ray photoemission experiment.

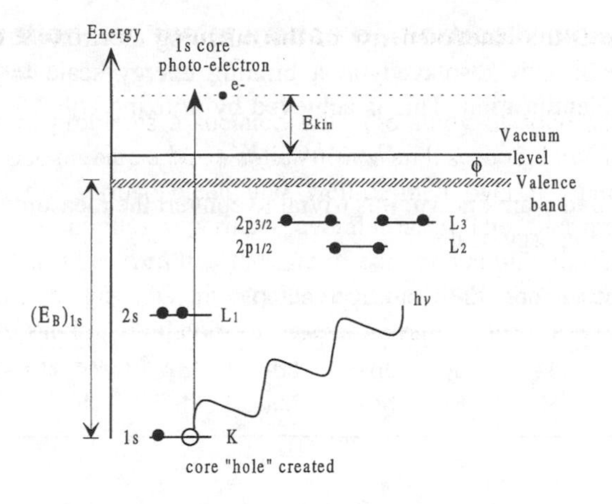

Fig. 2.3 X-ray excitation of a 1s core level.

If one assumes that the energy and spatial distribution of the electrons remaining after photoemission is exactly the same as in the initial state, one may simply equate the binding energy E_B with the negative orbital energy of the emitted electron:

$$E_B = -\varepsilon \ (\varepsilon = \text{orbital energy}) \quad (2.1a)$$

This approximation is called **Koopman's theorem**. However, in reality, the remaining electrons relax to a different energy state after photoemission and the core hole will influence the final state of the photoemitted electrons such that Koopman's theorem seldom applies. However this "Final State" shift is generally not more than a few electron volts allowing orbital identification to proceed.

spectroscopy, described later in Section 2.6. Applying the principle of energy conservation (Einstein's photoelectric equation), one may estimate the kinetic energy of emitted photoelectrons (E_{KIN}):

$$E_{\mathrm{KIN}} = \underbrace{h\nu}_{\text{Energy of photon}} - \underbrace{E_{\mathrm{B}} + \phi}_{\text{Binding energy of electron in solid}} \quad (2.1)$$

By convention the binding energy of a core level (E_{B}) is measured with respect to the highest occupied level of the solid, the Fermi level.* Figure 2.3 summarizes the XPS process. The total energy available to excite a core electron is clearly equal to the photon energy ($h\nu$). However, some of the photon energy must be consumed in overcoming the potential energy barrier, associated with attraction of the electron for the nucleus, ($E_{\mathrm{B}} + \phi$). The remaining energy is transformed into the kinetic energy of the photoemitted electrons. Hence it is clear that for a ***fixed*** photon energy, photoemission from an atom with well-defined core levels (of a particular binding energy) will produce photoelectrons with well-defined kinetic energies varying systematically from element to element. The higher the nuclear charge of an adatom the higher the binding energy of a given core level (e.g. E_{B} for an oxygen 1s level will be greater than that for a carbon 1s level). Table 2.1 summarizes the binding energies of elements of low atomic number.

*More correctly, the Fermi level of the spectrometer to which the sample is connected electrically.

The electron energy distribution [the number of electrons detected '$N(E)$' as a function of their kinetic energy] can be measured using an electrostatic energy analyser consisting of two electrically isolated concentric hemispheres with a potential difference between them, as shown in Fig. 2.4. The electrostatic field separates electrons by allowing only electrons of a chosen kinetic energy (the 'pass energy') through to the detector (continuous line). Electrons of kinetic energy less than the chosen 'pass energy' are attracted by the inner positive hemisphere and neutralized. The higher energy electrons hit the outer hemisphere and are also lost. The scanning of electron kinetic energies to produce a photoelectron spectrum is achieved by 'retarding' (slowing down) the electrons to the pass energy using a negative electrode (the retard plate). By changing the negative voltage on the retard plate, electrons with differing kinetic energies are allowed through the analyser to the detector.

While it is the kinetic energy of the outgoing electrons that is measured, spectra are usually displayed on a binding energy scale to allow ease of elemental identification. This is achieved by utilizing eqn 2.2

$$E_{\mathrm{B}} = h\nu - E_{\mathrm{kin}} - \phi \tag{2.2}$$

Here, if $h\nu$ and ϕ are known, it is trivial to convert the measured kinetic energy to a binding energy scale.

Table 2.1 Binding energies of elements of atomic number 1–38 in electron volts [1]

		$1s_{\frac{1}{2}}$ K	$2s_{\frac{1}{2}}$ L_I	$2p_{\frac{1}{2}}$ L_{II}	$2p_{\frac{3}{2}}$ L_{III}	$3s_{\frac{1}{2}}$ M_I	$3p_{\frac{1}{2}}$ M_{II}	$3p_{\frac{3}{2}}$ M_{III}	$3d_{\frac{3}{2}}$ M_{IV}	$3d_{\frac{5}{2}}$ M_V	$4s_{\frac{1}{2}}$ N_I	$4p_{\frac{1}{2}}$ N_{II}	$4p_{\frac{3}{2}}$ N_{III}
1	H	14											
2	He	25											
3	Li	55											
4	Be	111	†										
5	B	188	†	5*									
6	C	284	†	7*									
7	N	399	†	9*									
8	O	532	24	7*									
9	F	686	31	9*									
10	Ne	867	45	18*									
11	Na	1072	63	31*		1							
12	Mg	1305	89	52*		2							
13	A1	1560	118	74	73	1							
14	Si	1839	149	100	99	8	3						
15	P	2149	189	136	135	16	10						
16	S	2472	229	165	164	16	8						
17	Cl	2823	270	202	200	18	7						
18	Ar	3203	320	247	245	25	12						
19	K	3608	377	297	294	34	18*						
20	Ca	4038	438	350	347	44	26*		5*				
21	Sc	4493	500	407	402	54	32*		7*				
22	Ti	4965	564	461	455	59	34*		5*				
23	V	5465	628	520	513	66	38*		2*				
24	Cr	5989	695	584	575	74	43*		2*				
25	Mn	6539	769	652	641	84	49*		4*				
26	Fe	7114	846	723	710	95	56*		6*				
27	Co	7769	926	794	779	101	60*		3*				
28	Ni	8333	1008	872	855	112	68*		4*				
29	Cu	8979	1096	951	931	120	74*		2*				
30	Zn	9659	1194	1044	1021	137	87*		9*				
31	Ga	10 367	1298	1143	1116	158	107	103	18*				1
32	Ge	11 104	1413	1249	1217	181	129	122	29*				3
33	As	11 867	1527	1359	1323	204	147	141	41*				3
34	Se	12 658	1654	1476	1436	232	168	162	57*				6
35	Br	13 474	1782	1596	1550	257	189	182	70	69	27		5
36	Kr	14 326	1921	1727	1675	289	223	214	89*		24		11
37	Rb	15 200	2065	1864	1805	322	248	239	112	111	30	15	14
38	Sr	16 105	2216	2007	1940	358	280	269	135	133	38		20

* Average of two peaks (difficult to resolve).
† Take as a couple of eV (valence band).

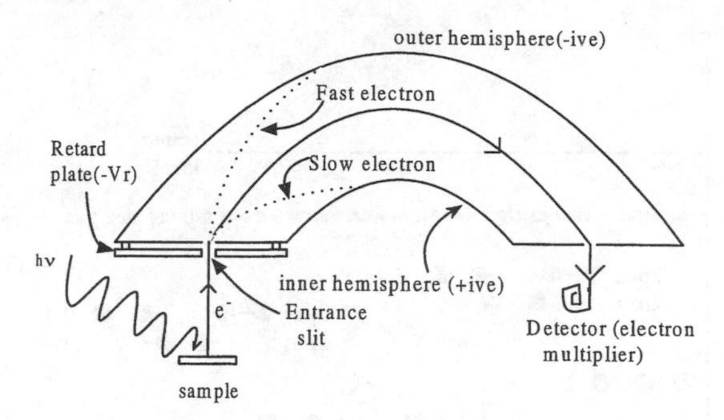

Fig. 2.4 Electrostatic energy analyser used in electron/ion spectroscopic analysis of surfaces.

Figure 2.5 illustrates an XP spectrum of a gold surface excited by Al-K_α radiation. The lowest binding energy peaks correspond to emission from the valence band (5d/6s electrons) followed by a series of XPS peaks of increasing binding energy. Note that:

(i) all core levels with orbital angular momentum quantum number ≥ 1 (p, d, f, ...) are split into doublets by spin–orbit coupling, with the higher angular momentum state at higher kinetic energy (lower binding energy); and

(ii) the background intensity at high binding energies increases owing to inelastically scattered electrons originating from parent XPS peaks (see Section 1.11 concerning 'secondary electrons').

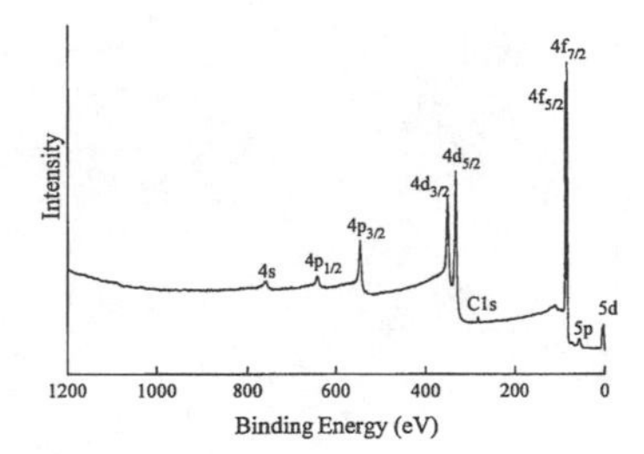

Fig. 2.5 The XP spectrum of a gold surface.

The spectrum also indicates the presence of carbon on the Au surface, as signified by the carbon 1s peak at 284 eV binding energy. The detection limit for surface impurities in XPS can, in favourable circumstances, be <1% of a monolayer.

Siegbahn and co-workers [2] (who pioneered the technique of XPS) showed that XPS is also a probe of the chemical environment or 'oxidation state' of surface species. Hence, XPS is often referred to as 'electron spectroscopy for chemical analysis' or ESCA.

The precise binding energy of the core levels of an atom or molecule will depend critically on the species to which it is bonded. Charge transfer may leave atoms with partial positive (or negative) charges, leading to a shift in core levels to higher (or lower) binding energies associated with increased (or decreased) Coulombic attraction between core electrons and the nucleus. Hence, atoms in a high formal oxidation state will yield XPS peaks at high binding energy relative to the same atom in a low oxidation state. The magnitude of this so called 'chemical shift' is dependent on the local environment surrounding the atom in question and can, in certain cases, be as large as 10 eV. Figure 2.6 illustrates the oxygen-induced shift in the core level binding energies of silicon. The insert shows the XPS of the silicon 2p region of a partially oxidized silicon sample [2]. The chemical shift between the peak at lower binding energy (Si^0) and the peak at higher binding energy (Si^{IV}) is approximately 4 eV. In addition, note that the intensity of the Si^{IV} species increases at grazing angles of emission since SiO_2 resides in a surface layer (see eqn 1.55).

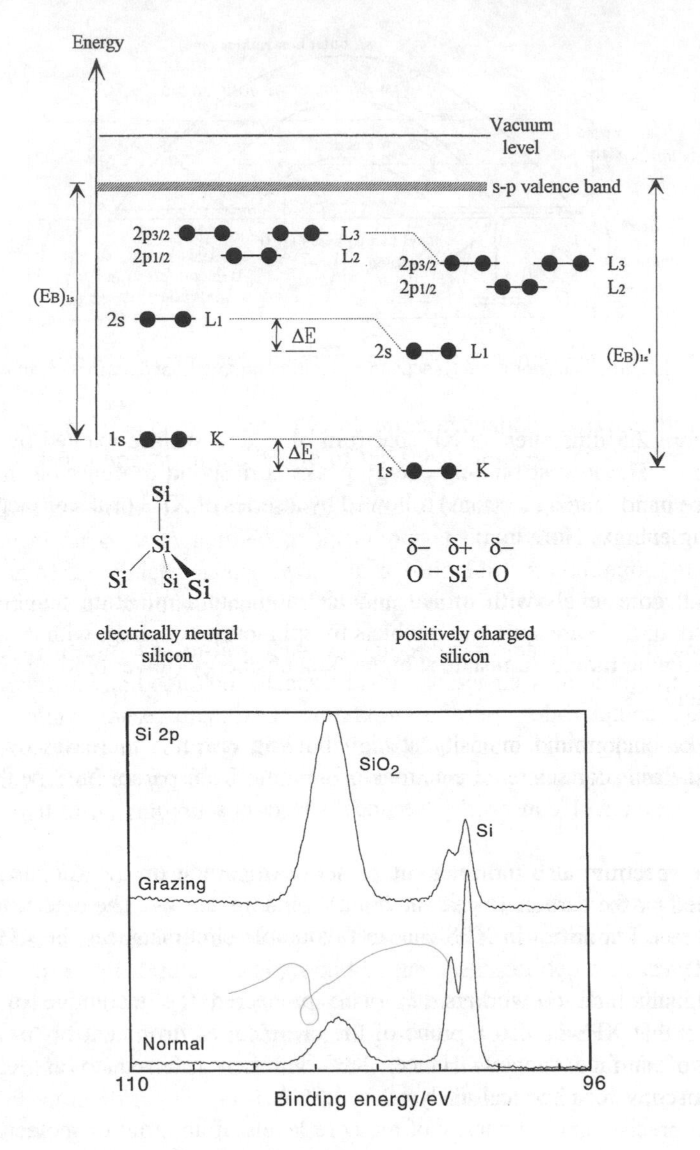

Fig. 2.6 Chemical shift of silicon core levels. Inset: XP spectra at grazing and normal emission from a thin layer of SiO_2 on Si. Adapted from ref. 2. Note the greater sensitivity to the surface SiO_2 at grazing emission.

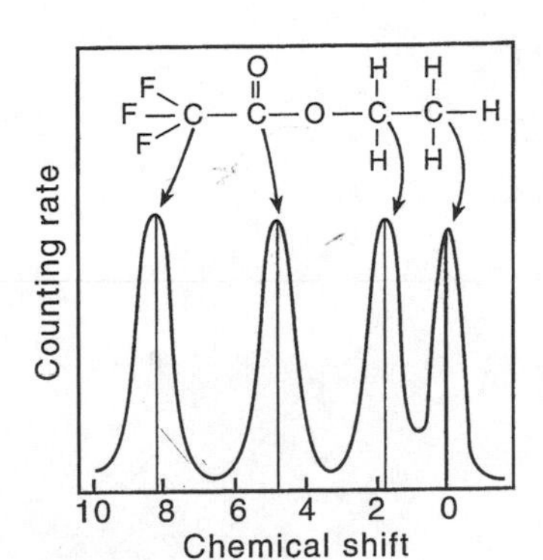

Fig. 2.7 The gas phase carbon 1s XP spectrum of ethyl trifluoroacetate showing four separate peaks corresponding to the four different chemical environments of carbon atoms in the molecule. Adapted from ref. 3.

While, in principle, a separate XPS peak should be observed for every chemically distinct atom (as shown in Fig. 2.7, which is a C 1s gas phase XPS spectrum of ethyl trifluoroacetate in which four carbon core levels are clearly resolved), this is often not the case in practice because the 'energy spread' of the incident radiation is often considerably larger than the chemical shift. Hence, peaks separated by small chemical shifts (of the order of tenths of an electron volt) remain unresolved.

The relative intensity of different XPS peaks will depend on a number of factors, including the concentration of atoms of an element in the selvedge, the probability of photoemission occurring for a particular core level (referred to as a photoemission cross-section), the IMFP (see fig. 1.24) of the photoemitted electron itself, and the efficiency of the spectrometer for detection of electrons as a function of kinetic energy (instrumental response). For XPS peaks of similar binding energy, both IMFP and instrumental factors may be ignored and if, for example, two elements, A and B, are distributed homogenously throughout the sampling depth, then relative concentrations may be obtained using the expression

$$\frac{C_A}{C_B} = \frac{I_A}{I_B} \bullet \frac{\sigma_B}{\sigma_A} \tag{2.3}$$

where C_A = atomic concentration of A; C_B = atomic concentration of B; I_A = XPS peak area of A core level; I_B = XPS peak area of B core level; σ_A = photoemission cross-section of core level of atom A; and σ_B = photoemission cross-section of core level of atom B.

Photoionization cross-sections have been tabulated for all the elements excited by both Al-and Mg-K_α radiation [1]. The only other case in which elemental compositional analysis can be carried out straightforwardly is that of a two-dimensional overlayer (see Question 6 in Chapter 3).

2.2 Auger electron spectroscopy (AES)

Auger electrons are named after their discover, Pierre Auger. They arise from an 'autoionization' process within an excited atom. Figure 2.8 illustrates the Auger effect. An incident photon (or electron) causes photoemission of a core electron (electron 1). The 'hole' (or electron vacancy) created in the core level by photoemission may be neutralized by an electron transition from an electron level of lower binding energy. This is called the 'down' electron (electron 2). The quantum of energy ΔE (equal to the difference in binding

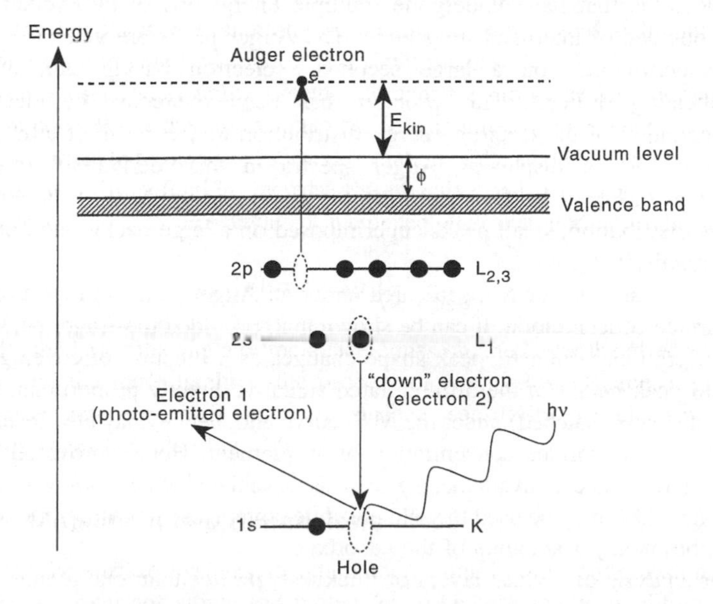

Fig. 2.8 Energetics of the Auger process.

energy between the core hole and the down electron) now becomes available and may either be removed from the atom as a photon (X-ray fluorescence) or transferred to a *third* electron, which can escape into the vacuum with a kinetic energy E_{KIN}. It is this third electron that is termed the Auger electron. The kinetic energy of the Auger electron is given by:

$$E_{kin} = \boxed{E_K - E_{L_1}} - \boxed{E_{L_{2,3}} - \phi} \quad (2.4)$$

Energy available to the Auger electron from 'down' electron.	Energy required to overcome barrier to emission from within atom

Auger peaks are usually assigned by three letters which specify (in X-ray notation) the levels from which the core hole, the 'down' electron, and the Auger electron originate. In the example shown in Fig. 2.8, the Auger transition is termed $KL_1L_{2,3}$.

The kinetic energy of the Auger electron, in contrast to photoemission is seen to be independent of the energy of the incident radiation giving rise to the initial core hole. The kinetic energy of the Auger electrons are characteristic solely of the binding energies of electrons *within the atom*. Hence, Auger electrons may be used for elemental identification.

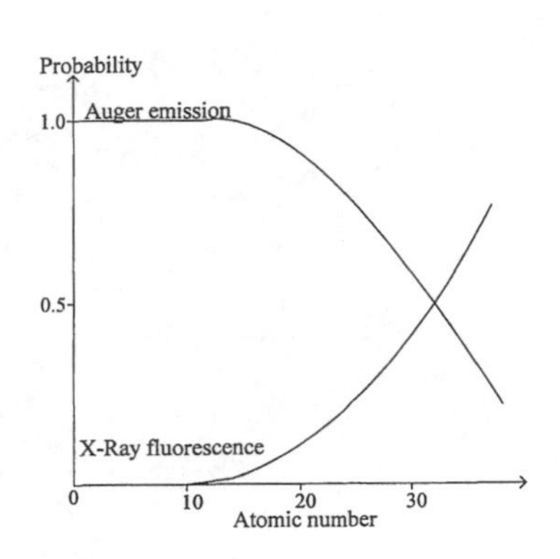

Fig. 2.9 Probability of Auger emission versus X-ray fluorescence as a function of atomic number.

All elements with three or more electrons (i.e. all elements other than H and He), give rise to a characteristic Auger spectrum, the complexity of which increases with atomic number owing to the greater number of possible transitions. Figure 2.9 illustrates the relative probabilities of X-ray fluorescence and Auger emission for a 1s core level as a function of atomic number. It is evident that Auger processes dominate for **elements of low atomic number**.

In most cases, initial core level excitation is performed with an electron beam. A typical energy distribution of the emitted electrons [in the form of the number of electrons with a given kinetic energy $N(E)$ against the kinetic energy] is shown in Fig. 1.22. Most of the emitted electrons are 'secondaries', i.e. electrons that have undergone multiple energy losses by excitation of plasmons and/or interband transitions. The Auger peaks are generally small and superimposed on a large secondary electron background, making identification difficult. This problem has been overcome by electronic differentiation of the electron energy distribution to yield a $dN(E)/dE$ curve. The principle of displaying Auger spectra in this 'derivative' mode is illustrated in Fig. 2.10. By measuring the change in gradient of the electron energy distribution, small peaks superimposed on a large background may be more readily detected.

As was the case for XPS, the area under an Auger peak is proportional to the surface concentration. It can be shown that, provided the Auger peak does not undergo any dramatic peak shape changes as a function of coverage, the peak to peak height of the differentiated signal is directly proportional to the integrated area (hatched) under the $N(E)$ curve and, hence, may also be used as a probe of the surface concentration of an element. Hence, provided that a point of reference is available, i.e. a peak associated with a *known* surface coverage, AES may be used to yield absolute coverages, at least for monolayer and submonolayer amounts of the adsorbate.

For analysis of surface layers of thickness *greater* than one atomic layer, Auger electrons from atoms in the first layer must pass through second layer

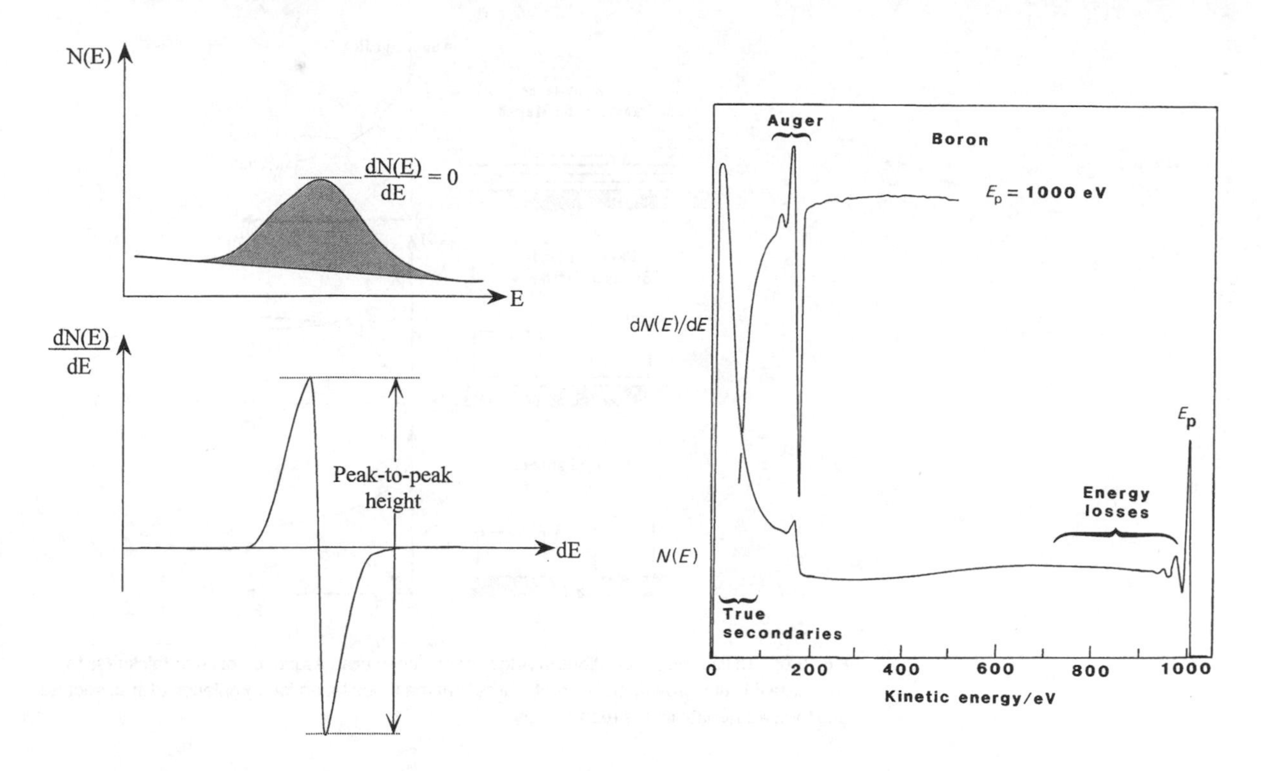

Fig. 2.10 Auger peak $\{N(E)\}$ and differentiated Auger peak $\{dN(E)\}/dE$. The Auger spectrum of boron in both $N(E)$ and $dN(E)/dE$ modes. Note the much greater sensitivity to Auger peaks in the differentiated signal. Incident kinetic energy of primary electron beam $(E_p) = 1000$eV.

material to reach the detector (Fig. 2.11). Inelastic energy losses may occur en route to the detector, thus leading to a smaller contribution from the first monolayer compared with the outermost layer. For layers of thickness, 3, 4, 5, 6, etc. atoms, the contribution of the first layer becomes subsequently less and less. This effect of 'signal damping' with increasing coverage has found extensive use in the study of the earliest stages of crystal growth ('epitaxy'). The morphology of thin films is increasingly important to a range of technologies, including the formation of metal–semiconductor junctions in electronic devices and the production of metallic magnetic multilayers for data storage. The properties of these metallic films are highly dependent on the growth mechanism of the thin layer. Because Auger electron emission is modulated strongly by the thickness of an adsorbate layer, by monitoring the intensity of an adsorbate Auger peak as a function of deposition time at constant flux, the growth mechanism of the adsorbate film can be characterized.

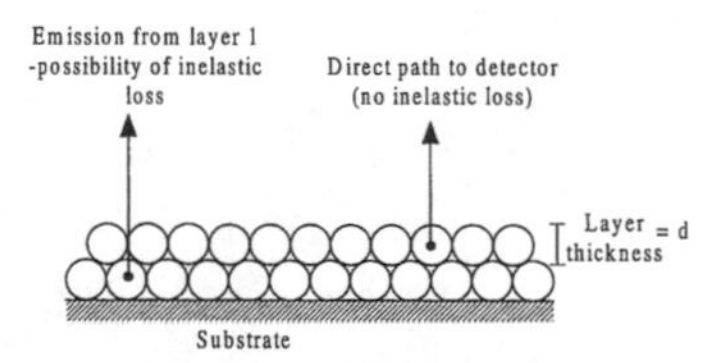

Fig. 2.11 Difference in likelihood of inelastic energy loss for surface and second layer adsorbate atoms.

Using Table 2.1 and eqn 2.4, estimate the theoretical value of the kinetic energy for the KLL Auger transition for boron. Does this value agree with the value in fig 2.10?

Figure 2.12 illustrates the three main types of growth mechanism and their accompanying Auger signal versus deposition time (AS–*t*) plots. In the first mechanism the film grows layer by layer (Frank–van der Merwe growth), in which growth of a new atomic layer does not start until the preceding layer is complete. In mechanism two, growth occurs in the form of three-dimensional crystallites (Volmer–Weber), i.e. although bare patches of the substrate are available, the adsorbate prefers to form multiple atomic layers. Finally, the third mechanism is an intermediate situation, between the extremes of

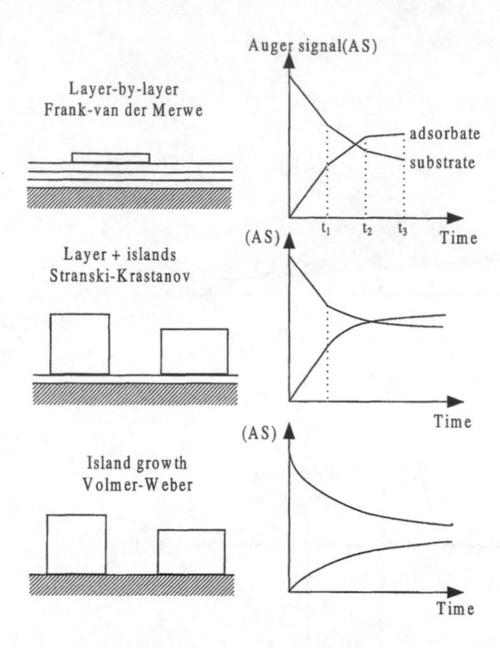

Fig. 2.12 Utilization of variations in Auger peak intensity as a function of layer thickness to elucidate thin film growth. t_1, t_2 and t_3 represent the times taken for completion of first, second and third adsorbate layers respectively.

mechanisms one and two, in which the first one (or few) atomic layers grow in a layer by layer fashion. When a critical film thickness is reached subsequent growth occurs in the form of three-dimensional crystallites (Stranski–Krastanov mechanism). The form of the Auger signal versus time plots is different in each of these three cases. The layer by layer mode is characterized by a series of linear segments of differing gradient. The 'break points' (where each linear segment intersects) correspond to the completion of each monolayer. The attenuation in the Auger electron signal from the substrate, (I_s^n) at normal emission, associated with n atomic adsorbate layers each of thickness d_a and an IMFP of λ_s is given by

$$I_s^n = I_s^0 \exp(-nd_a/\lambda_s) \tag{2.5}$$

where I_s^0 is the Auger intensity from the *clean* substrate. Similarly, the increase in intensity of the adsorbate Auger peak, (I_a^n), associated with n atomic adsorbate layers of thickness d_a and IMFP of λ_a is given by

$$I_a^n = I_a^\infty[1 - \exp(-nd_a/\lambda_a)] \tag{2.6}$$

In the case of the Volmer–Weber growth mechanism, the AS–t plot will consist of a smooth decay of the substrate Auger signal accompanied by a smooth increase of the overlayer Auger peak intensity. The exact rate of decay of the substrate peak (increase of the overlayer signal) will depend on the precise shape of the clusters formed on the surface, but will always be slower than that for the layer by layer mechanism. The Stranski–Krastanov mode leads to an intermediate form of plot with one or more linear segments (break points) as the initial deposit adsorbs, followed by the more gradual monotonic change resulting from subsequent bulk crystallite growth.

A second application of Auger spectroscopy, termed scanning Auger microscopy (SAM), utilizes the fact that high energy electron beams can be focused down to spot sizes of only several hundred Ångstroms (below a thousandth of a millimetre!). This means that an Auger spectrum can be collected from a well-defined microscopic region on the sample surface. By scanning (rastering) the electron beam across the surface whilst simultaneously monitoring a particular Auger peak, an 'image' of the lateral variation in elemental concentration across the surface can be built up. Elemental composition as a function of depth into a sample may also be obtained by combining argon ion bombardment (see section 1.9) with AES. This technique is known as 'depth profiling'. In depth profiling, the Ar ions etch away layers of material at a rate typically around several monolayers per minute, allowing an incident electron beam focused on the Ar^+-bombarded area of the sample to excite Auger emission as a function of time and hence depth (assuming a constant rate of layer removal). Provided the sputter rate can be calibrated by the use of films of known thickness (standards), a plot of Auger signal (peak to peak height) versus depth is obtained. This is termed a 'depth profile'. While problems remain because of Ar^+-induced mixing at atomically sharp interfaces, depth profiling is one of the few ways of monitoring elemental concentrations to depths between hundreds and thousands of Ångstroms. This is extremely important in the analysis of thin films and coatings.

2.3 Low energy electron diffraction (LEED)

In LEED, incident electrons, elastically back scattered (without energy loss) from a surface, are analysed in the energy range 20–1000 eV. As illustrated in Fig. 1.24, electrons in this energy range possess inelastic mean free paths of between ~5 and 20 Å and, therefore, may only travel a few atomic layers into the surface. Electrons in this energy range are excellent probes of surface structure because they possess de Broglie wavelengths of the same order of magnitude as the interatomic spacing between atoms/molecules at surfaces and, hence, may undergo **diffraction** if the atoms in the surface are arranged periodically.

The wavelength of electrons may be estimated from the modified de Broglie equation:

$$\lambda(\text{Å}) = \left(\frac{150.6}{E(\text{eV})}\right)^{\frac{1}{2}} \tag{2.7}$$

yielding de Broglie wavelengths in the range 2.74–0.388 Å for kinetic energies between 20 and 1000 eV.

Figure 2.13 illustrates schematically the experimental apparatus necessary to perform a LEED experiment. In essence, the LEED apparatus ensures that only those electrons of kinetic energy E_p (the primary beam energy) reach the phosphor screen (detector). If an ordered surface array is prepared, discrete electron beams emerge from the substrate whose spatial distribution reflects the **symmetry** of the ordered array (*cf.* X-ray diffraction). The LEED experiment itself may be summarized as follows.

A monochromatic electron beam whose energy (E_p) can be varied (typically in the range 0–1000 eV) is generated by an electron gun. The

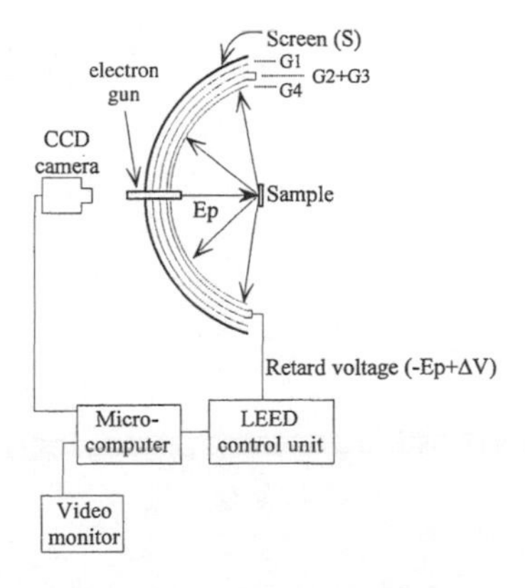

Fig. 2.13 The LEED apparatus. Electrons of kinetic energy E_p are directed at the sample from an electron gun. The various grids G1–G4 ensure that only those electrons elastically scattered from the sample reach the phosphor screen.

beam is incident upon a sample that must be an electrical conductor connected to earth to prevent charging. After undergoing diffraction, electrons back-scattered from the periodic surface travel towards a series of concentric meshes or grids (G1–4). The outer grid (G4) nearest to the sample is earthed to ensure that the electrons travel in a 'field free' region, as is the inner grid (G1). The earthing of G1 screens out the high voltages placed on the phosphor screen (S). The inner pair of grids (G2 and G3) serve as a cut-off filter and are held at a negative potential $(- E_p + \Delta V)$, where ΔV is typically in the range 0–10 V. This ensures that only elastically scattered electrons reach the detector, S. S is biased at a high positive voltage (~6 keV) to accelerate the transmitted electrons to a sufficient kinetic energy to cause light emission from the coated fluorescent glass screen. The diffracted electrons give rise to a pattern consisting of bright spots on a dark background, which reflect the symmetry and crystalline order of the surface. The LEED pattern may either be viewed by eye or monitored with a video camera, if quantitative intensity measurements are required.

A range of information is available from a LEED pattern.

(i) From the position of the diffracted beams, the two-dimensional periodicity of the surface unit cell may be deduced, along with variations in the unit cell size induced by adsorption.

(ii) From the variation of spot intensities with beam energy, the complete surface geometry, including bond lengths and angles, can be obtained.

In order to explain the LEED process, we shall first consider the simplest possible case: diffraction from a one-dimensional periodic array. Figure 2.14 illustrates electrons scattering at an angle θ_a from atoms in a one-dimensional chain (lattice constant $= a$). For constructive interference between scattered electron waves, the path length difference must be equal to an integral number of wavelengths. By simple geometry, the path length difference (Δ_a) is

$$\Delta_a = a \sin \theta_a \tag{2.8}$$

and for constructive interference

$$\Delta_a = n\lambda \tag{2.9}$$

where λ is the de Broglie wavelength of the electron and is *constant* at a fixed incident electron kinetic energy. Combining eqns 2.8 and 2.9

$$n\lambda = a \sin \theta_a \tag{2.10}$$

(where n can take values 0, ± 1, ± 2, ± 3,...)
Re-arrangement yields

$$\sin \theta_\mathrm{a} = \frac{n\lambda}{a} \tag{2.11}$$

Hence, for a fixed wavelength, λ, and lattice spacing, a, only well-defined values of θ_a are allowed for which constructive interference will be observed corresponding to integer values of n. This means that **discrete diffracted beams** will be seen at particular angles.

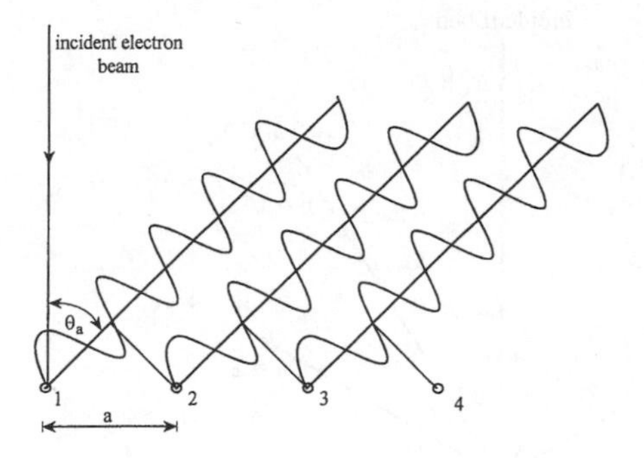

Fig. 2.14 Diffraction from a one-dimensional array.

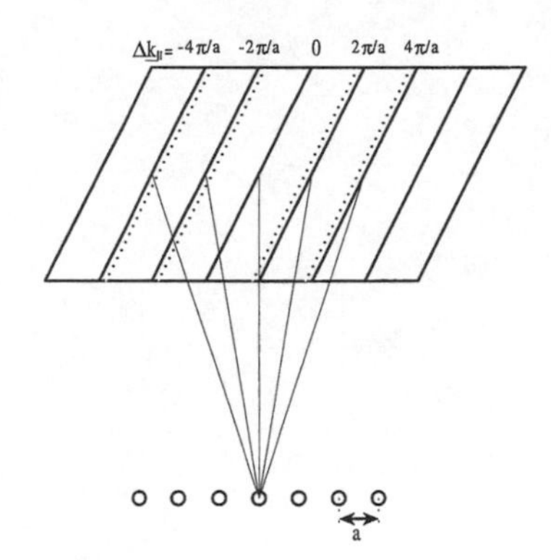

Fig. 2.15 Diffraction pattern observed from a one-dimensional array. Dotted lines represent change in diffraction pattern caused by increasing kinetic energy of primary electron beam (Decreasing λ).

Figure 2.15 illustrates the diffraction pattern obtained from a one-dimensionally periodic solid. The pattern consists of a series of equally spaced lines **perpendicular** to the one-dimensional lattice in which the atom spacing is **inversely** related to the periodic spacing in the diffraction pattern. Equation 2.11 indicates that for fixed λ, as 'a' increases, $\sin \theta_a$, and, hence θ_a, will decrease, leading to diffracted beams becoming more narrowly spaced. In a similar manner, for fixed lattice constant a, if the electron wavelength is decreased (kinetic energy increased), θ will decrease and the diffracted beams will move closer together as illustrated by the dotted lines in Fig. 2.15.

An alternative way of representing the condition for diffraction is in terms of 'electron wavevectors' and so called 'reciprocal lattice vectors'.

The magnitude of the incident wavevector of an electron ($\boldsymbol{k}_0$) is defined as

$$|\boldsymbol{k}_0| = \frac{2\pi}{\lambda} \tag{2.12}$$

and is a measure of an electron's momentum. This may be demonstrated simply by recalling the de Broglie relationship

$$\lambda = \frac{h}{mv} \tag{2.13}$$

By combining eqns 2.12 and 2.13

$$|\boldsymbol{k}_0| = \frac{2\pi}{h}(mv) \tag{2.14}$$

$$mv = \text{momentum}$$

Substitution of eqn 2.11 into eqn 2.12 and eliminating "λ" then gives

$$|\boldsymbol{k}_0| \sin \theta_a = \left(\frac{2\pi}{a}\right)n \tag{2.15}$$

But $|\boldsymbol{k}_0|$ sin θ_a is the component of momentum **parallel** to the surface of the incident electron ($\boldsymbol{k}_{\|}$) (see Fig. 2.16). Furthermore, it is evident from eqn 2.15 that parallel momentum may only be exchanged with the surface in quantized units of '$2\pi/a$'. $\{2\pi/a\}$ is the magnitude of the one-dimensional **reciprocal lattice vector**. Since, from Fig. 2.17, incident electrons that are normal to the

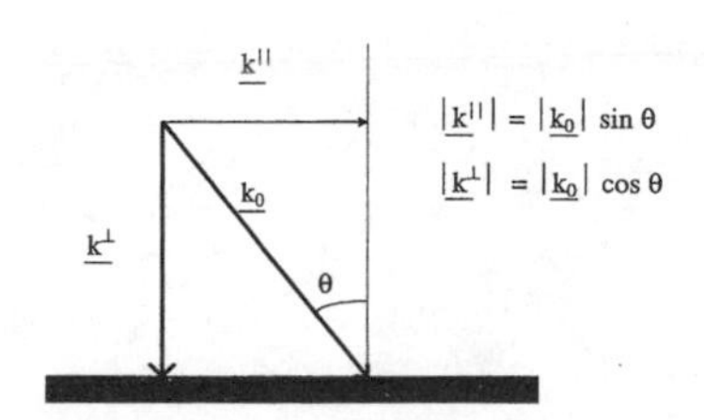

Fig. 2.16 Resolution into parallel and perpendicular components of an electron with incident wave vector $\boldsymbol{k}_0$.

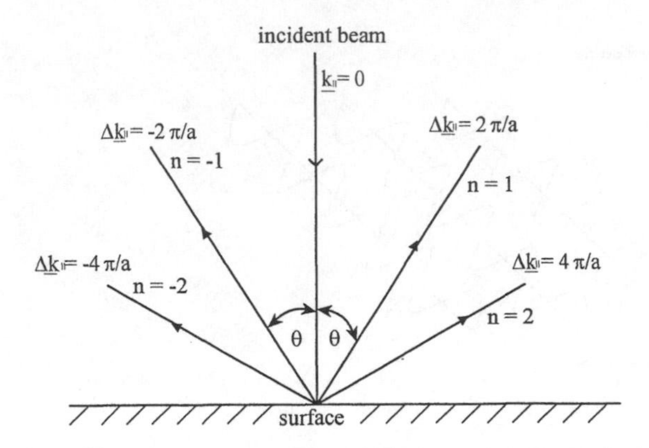

Fig. 2.17 Change in scattering direction of diffracted electrons associated with change in parallel momentum $\Delta \boldsymbol{k}_{\|}$. Note that $\Delta \boldsymbol{k}_{\|}$ may only take integral values of the reciprocal lattice vector 2π/a.

one-dimensional array initially have no component parallel to the chain, for diffracted beams to arise (for the electrons to undergo a change in direction), the electron must exchange parallel momentum with the one-dimensional lattice (i.e. momentum is **conserved**)

$$\Delta \boldsymbol{k}_{\|} = |\boldsymbol{k}_0| \sin\theta_a = \left(\frac{2\pi}{a}\right)n \tag{2.16}$$

where $\Delta \boldsymbol{k}_{\|}$ represents the change in parallel momentum in quantized units of (2π/a).

As mentioned earlier, in the case of a one-dimensional array, the diffraction pattern consists of a set of parallel lines. If we now introduce periodicity in a second (orthogonal) direction in which the repeat distance is *b*, the condition for constructive interference may be derived analogous to equation 2.11

$$\sin\theta_b = \frac{m\lambda}{b} \tag{2.17}$$

leading to a second set of diffracted beams perpendicular to the lattice direction and whose spacing is inversely related to the lattice spacing, *b*, as shown in Fig. 2.18. The periodicity in the second dimension restricts parallel momentum exchange to

$$\Delta \boldsymbol{k}_{\|} = |\boldsymbol{k}_0| \sin\theta_b = \left(\frac{2\pi}{b}\right)m \tag{2.18}$$

where *m* may take values of 0, ± 1, ± 2, ± 3...

Both eqns 2.16 and 2.18 must be satisfied *simultaneously* for diffraction to be observed from a two-dimensional array. Thus two-dimensional diffraction is allowed *only* at the **intersection** of the one-dimensional reciprocal lattice rods generated in the *a* and *b* directions, respectively, and the LEED pattern consists of a series of diffraction spots or 'beams' corresponding to these points of intersection. In this case, the exchange of parallel momentum is restricted to a **two-dimensional reciprocal lattice vector** (***G***) where

$$\boldsymbol{G} = \Delta \boldsymbol{k}_{\|} = n\frac{2\pi}{a} + m\frac{2\pi}{b} \tag{2.19}$$

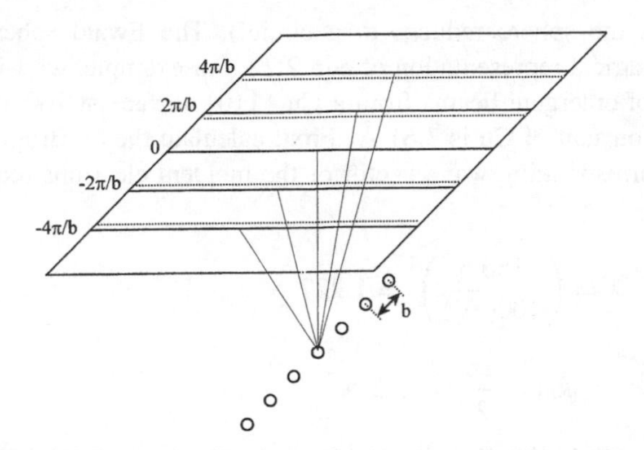

Fig. 2.18 Diffraction pattern observed from a one-dimensional array in a direction perpendicular to that in Fig. 2.15. Dotted lines represent change in diffraction pattern caused by increasing kinetic energy of primary electron beam (decreasing λ).

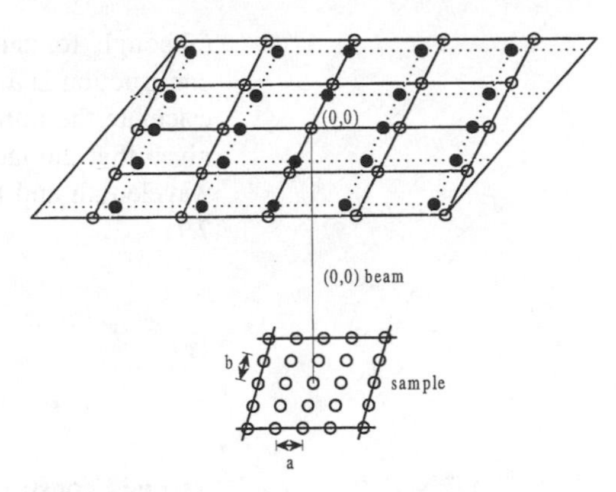

Fig. 2.19 Diffraction pattern observed from two-dimensional array. Diffraction spots occur when $\Delta \boldsymbol{k}_{\parallel}$ corresponds to a two-dimensional reciprocal lattice vector, i.e. when the diffraction lines in Figs 2.15 and 2.18 coincide. Note how the pattern "contracts" towards the (0, 0) beam as λ decreases (filled circles).

The dotted lines in Fig. 2.19 illustrate the effect of increasing the beam energy (decreasing the de Broglie wavelength) in that the diffracted beams move inwards towards the '**origin**' of the diffraction pattern, i.e. towards the diffracted beam, which undergoes **no parallel momentum change** ($n = m = 0$) termed the (0,0) beam. Other diffracted beams are labelled by the values (n,m) that characterize their parallel momentum transfer.

Although the conceptualization of two-dimensional diffraction may readily be understood in terms of Bragg-like equations (eqns 2.11 and 2.17), the **machinery** best suited for predicting the distribution of diffracted beams from a surface in terms of their number and direction at a given primary beam energy and angle of incidence is more readily understood in terms of reciprocal lattice (G vector) analysis.

Every real space lattice will generate an associated 'reciprocal lattice' upon diffraction, constructed using the following set of rules:

$$G = n\boldsymbol{a}^* + m\boldsymbol{b}^* \quad (n \text{ and } m \text{ are integers}) \tag{2.20}$$

$$|\boldsymbol{a}^*| = 2\pi/|\boldsymbol{a}|;\ |\boldsymbol{b}^*| = 2\pi/|\boldsymbol{b}^*|;\ \boldsymbol{a}.\boldsymbol{b}^* = \boldsymbol{a}^*.\boldsymbol{b} = 0 \tag{2.21}$$

*refers to reciprocal lattice vector

where $\boldsymbol{a}$ and $\boldsymbol{b}$ are the elementary vectors of the surface two-dimensional unit cell, and $\boldsymbol{a}^*$ and $\boldsymbol{b}^*$ are the elementary vectors of the corresponding reciprocal lattice. These equations state, in mathematical terms, that a large (small) distance in real space becomes a small (large) distance in reciprocal space and, in addition, that $\boldsymbol{a}$ and $\boldsymbol{b}$ are perpendicular to the direction of $\boldsymbol{a}^*$ and $\boldsymbol{b}^*$, respectively (recall Figs. 2.15 and 2.18). The condition for diffraction is then given by

$$\boldsymbol{k}_0^{\parallel} = \boldsymbol{k}_s^{\parallel} \pm \boldsymbol{G} \tag{2.22}$$

where $\boldsymbol{k}_s^{\parallel}$ is the parallel component of the **scattered** electron.

The consequence of conservation of parallel momentum $\boldsymbol{k}_0^{\parallel} = \boldsymbol{k}_s^{\parallel} \pm G$ means that, although energy is conserved in the diffraction process, if a G vector is exchanged with the surface, the incident electron must undergo a change in direction in order to conserve momentum.

A simple method of finding the number of diffracted beams emerging from a surface at a given energy is to use the **Ewald sphere construction** (although for surfaces the sphere reduces to a circle!). The Ewald sphere construction is a geometrical representation of eqn 2.22. For example, we will calculate the number of emergent beams from a Cu {110} surface at 100 eV given that the lattice constant of Cu is 3.61 Å. First, calculate the de Broglie wavelength and the corresponding wave vector of the incident electrons (eqn 2.7)

$$\lambda = \left(\frac{150.6}{100(\mathrm{eV})}\right)^{\frac{1}{2}} = 1.227\ \text{Å}$$

$$|\boldsymbol{k}_0| = \frac{2\pi}{\lambda} = 5.12\ \text{Å}^{-1}$$

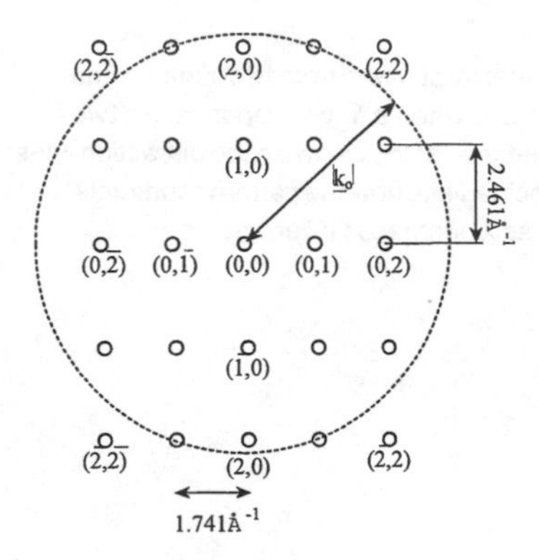

Fig. 2.20 Ewald sphere construction, a geometrical solution of eqn 2.22. For a Cu(110) surface (see fig.1.14), $|\boldsymbol{a}| = 3.61\text{Å}/\sqrt{2} = 2.55\text{Å}$
$\therefore |\boldsymbol{a}^*| = 2\pi/2.55\text{Å} = 2.461\text{Å}^{-1}$
$|\boldsymbol{b}| = 3.61\text{Å}$
$\therefore |\boldsymbol{b}^*| = 2\pi/3.61\text{Å} = 1.741\text{Å}^{-1}$

Second, construct, to scale, the two-dimensional reciprocal lattice of the surface using the equations listed under eqn 2.21, noting that $\boldsymbol{a}$ is perpendicular to $\boldsymbol{a}^*$ and $\boldsymbol{b}$ is perpendicular to $\boldsymbol{b}^*$. Third, choose a reciprocal lattice point as the origin (0, 0). Finally, draw, to scale, a circle radius $|\boldsymbol{k}_0|$ centred at the origin. The total number of diffracted beams emerging from the surface is simply given by the number of reciprocal lattice points contained within the circle. From Fig. 2.20 it is seen that, at 100 eV, the number of diffracted beams is 21.

If one wishes to calculate the angle made by a diffracted beam with a particular real space direction, one simply needs to perform the Ewald circle construction once again. In this case, the reciprocal lattice will consist of a series of lines rather than an array of dots, since, by defining a particular direction, we generate a one-dimensional array and a corresponding set of diffraction *lines* in contrast to the two-dimensional array which produced a two-dimensional net of lattice points—see Figs 2.19 and 2.21. Figure 2.21 shows the Ewald circle construction for electrons at normal incidence to the Cu(110) surface for scattering in the $[1\bar{1}0]$ direction. Note that the point at which the circle generated by $|\boldsymbol{k}_0|$ cuts the reciprocal lattice rods satisfies eqn 2.22 and hence defines the direction θ at which the diffracted beams will emerge. Those beams pointing into the vacuum will be detected. However, those beams pointing into the solid will not reach the detector. By simple

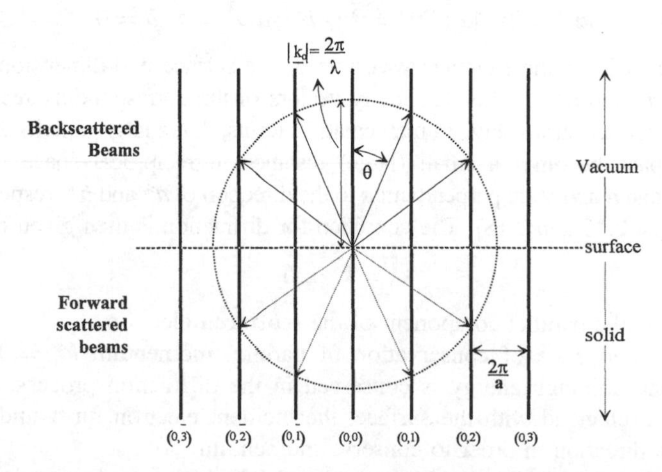

Fig. 2.21 Ewald sphere construction for diffraction into a particular surface direction. Note that the points in Fig. 2.20 become rods in Fig. 2.21. $|\boldsymbol{a}| = 3.61\text{Å}$ for $[1\bar{1}0]$ direction.

geometry, in the case of the (0, 1) beam

$$\sin\theta = \frac{2\pi/a}{2\pi/\lambda} = \frac{\lambda}{a} = \frac{1.227}{3.61} \therefore \theta = 19.9^\circ$$

It is quite often found that adsorption on to single crystals produces overlayers that are ordered and have a periodicity larger than that of the substrate unit cell, and, therefore, give rise to additional diffraction beams. The periodicity of the overlayer in real space may be deduced from observation of the LEED pattern.

Again, this is best illustrated with an example. The adsorption of 0.25 ML of atomic oxygen on Cu(001) gives rise to the change in LEED pattern illustrated in Fig. 2.22. In order to derive the unit cell of the overlayer in real space, one needs to perform a 'Fourier transform' of the reciprocal lattice (the LEED pattern). Straightforward manipulation of matrices associated with the overlayer mesh in reciprocal and real space may also achieve this goal.

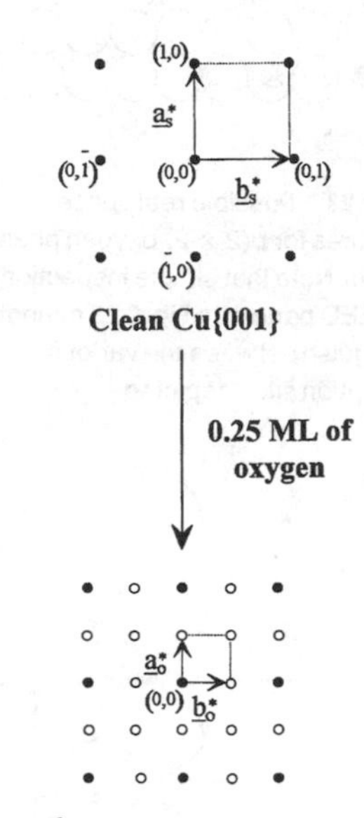

Fig. 2.22 Change in clean surface LEED pattern when 0.25 ML of oxygen is adsorbed on Cu(100).

(i) Write the reciprocal unit cell vectors of the overlayer in terms of their reciprocal space substrate counterparts (Fig. 2.22):

$$\begin{aligned} \boldsymbol{a}_0^* &= G_{11}^* \boldsymbol{a}_s^* + G_{12}^* \boldsymbol{b}_s^* & & \boldsymbol{a}_0^* = 1/2\boldsymbol{a}_s^* + 0\boldsymbol{b}_s^* \\ & & \rightarrow & \\ \boldsymbol{b}_0^* &= G_{21}^* \boldsymbol{a}_s^* + G_{22}^* \boldsymbol{b}_s^* & & \boldsymbol{b}_0^* = 0\boldsymbol{a}_s^* + 1/2\boldsymbol{b}_s^* \end{aligned} \tag{2.23}$$

(ii) Form the matrix

$$G^* = \begin{bmatrix} G_{11}^* & G_{12}^* \\ G_{21}^* & G_{22}^* \end{bmatrix} = \begin{bmatrix} \frac{1}{2} & 0 \\ 0 & \frac{1}{2} \end{bmatrix} \tag{2.24}$$

(iii) To convert from reciprocal space (diffraction pattern) to real space we take the 'inverse transpose' of the matrix G^*.

$$\begin{aligned} G &= \left([G^*]^{-1}\right)^{\mathrm{t}} = \frac{1}{\det G^*}\begin{bmatrix} G_{22}^* & -G_{21}^* \\ -G_{12}^* & G_{11}^* \end{bmatrix} \\ G &= \frac{1}{\frac{1}{4} - 0}\begin{bmatrix} \frac{1}{2} & 0 \\ 0 & \frac{1}{2} \end{bmatrix} = \begin{bmatrix} 2 & 0 \\ 0 & 2 \end{bmatrix} \end{aligned} \tag{2.25}$$

where $\det G^*$ is the determinant of the matrix G^* and is obtained by cross-multiplication.

$$\det G^* = (G_{22}^* . G_{11}^*) - (G_{21}^* . G_{12}^*) \tag{2.26}$$

The matrix G yields the real space overlayer vectors ($\boldsymbol{a}_0$ and $\boldsymbol{b}_0$) in terms of the real space substrate unit vectors ($\boldsymbol{a}_s$ and $\boldsymbol{b}_s$), i.e.

$$\begin{pmatrix} \boldsymbol{a}_0 \\ \boldsymbol{b}_0 \end{pmatrix} = G \begin{pmatrix} \boldsymbol{a}_s \\ \boldsymbol{b}_s \end{pmatrix} \tag{2.27}$$

Hence

$$\begin{aligned} \boldsymbol{a}_0 &= 2\boldsymbol{a}_s + 0\boldsymbol{b}_s \\ \boldsymbol{b}_0 &= 0\boldsymbol{a}_s + 2\boldsymbol{b}_s \end{aligned}$$

(iv) Finally, draw the real space overlayer mesh ($\boldsymbol{a}_0$ and $\boldsymbol{b}_0$) in terms of the substrate mesh ($\boldsymbol{a}_s$ and $\boldsymbol{b}_s$) as shown in Fig. 2.23.

To go from a real space structure to its diffraction pattern, simply follow the same procedure as above except for G read G^*, and for reciprocal lattice

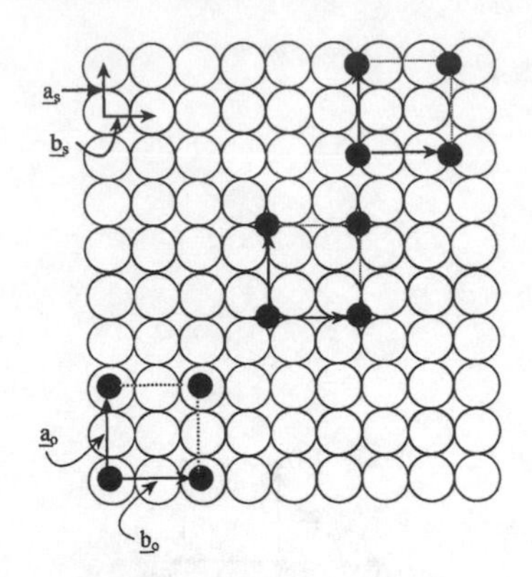

Fig. 2.23 Possible real space structures for p(2 × 2) oxygen phase on Cu(100). Note that simple inspection of the LEED pattern in Fig. 2.22 cannot distinguish between the various adsorption sites depicted.

vectors use real space vectors (see Question 7 of Chapter 3). Hence, in matrix notation, the LEED pattern for 0.25 ML of oxygen on copper (100) corresponds to the real space structure

$$\text{Cu (100)} - \begin{bmatrix} 2 & 0 \\ 0 & 2 \end{bmatrix} - \text{O} - (0.25\ \text{ML})$$

and in the Wood's notation, the same structure may be termed

$$\text{Cu (100)} - \text{p}(2 \times 2) - \text{O} - (0.25\ \text{ML})$$

While such a procedure may allow an **exact** definition of the two-dimensional periodicity of the overlayer relative to the substrate, it says nothing about the actual adsorption site (atop, bridge, fourfold hollow). For example, the unit cell shown in the bottom left of Fig. 2.23 may be shifted by a combined translation of $1/2\boldsymbol{a}_s + 1/2\boldsymbol{b}_s$, leading to placement of all overlayer atoms in fourfold hollow sites. Both overlayer registries (bridge and fourfold hollow) will produce the same LEED pattern. Simple inspection of the diffraction pattern only allows information to be obtained on the relative sizes of the substrate and overlayer unit cells, but no discrimination between adsorption sites or details of surface bond lengths and angles.

However, quantitative structural information, such as the type of adsorption site occupied and the bond lengths and bond angles of adsorbates, is contained within the variation of the diffracted beam intensities with energy. The simplest method of illustrating how more detailed structural information is contained within such measurements is to imagine diffraction without parallel momentum transfer (normal reflection) from an ideal surface with an interlayer spacing d (Fig. 2.24).

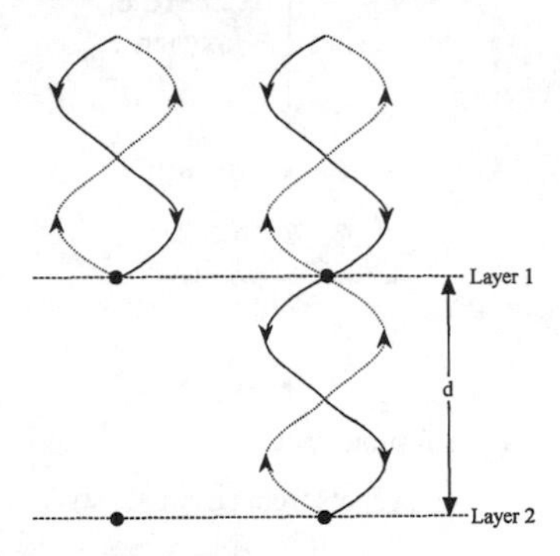

Fig. 2.24 Diffraction with zero momentum transfer parallel to the surface, interlayer spacing *d*. The bold line indicates incident electron waves, the dotted line reflected outgoing waves.

For constructive interference between electrons diffracting from the first and second layers we must set the path length difference ($2d$) equal to a whole number of wavelengths ($n\lambda$)

$$n\lambda = 2d \qquad (2.28)$$
$$n = 1, 2, 3, 4,$$

Note that the path length difference is now $2d$ since an electron back-scattered from layer 2 must travel a further distance d prior to scattering and an additional distance, d, on its backward journey en route to the detector, relative to scattering from layer 1.

Upon squaring eqn 2.28

$$n^2\lambda^2 = 4d^2 \qquad (2.28a)$$

Squaring eqn 2.7:

(λ in Å, E in eV)

$$\lambda^2 = \frac{150.6}{E} \qquad (2.29)$$

Hence, combining eqns 2.28a and 2.29 gives

$$E = \frac{150.6\, n^2}{4d^2} \qquad (2.30)$$

which corresponds to the energies at which we may, therefore, expect 'Bragg-like' reflections.

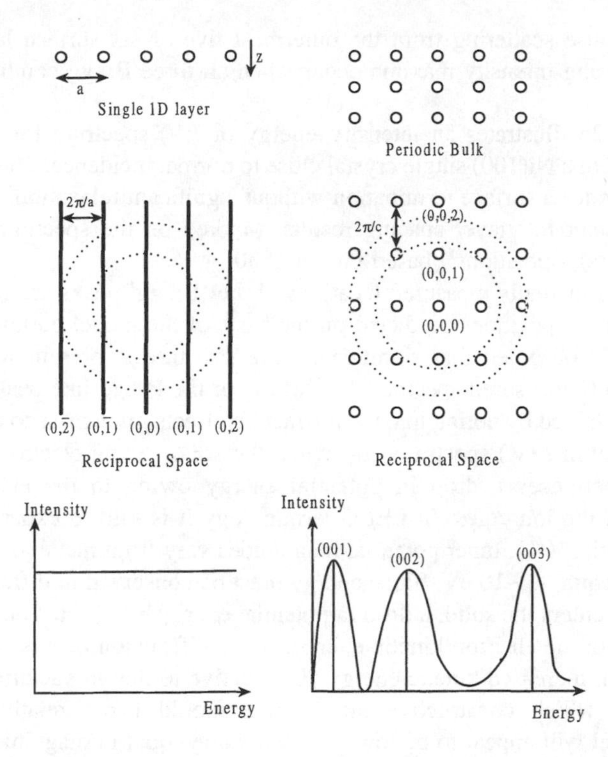

Fig. 2.25 Ewald sphere construction for Bragg diffraction from two- and three-dimensional crystals. Note that the variation in diffraction intensity with energy is constant for diffraction from a two-dimensional layer, but that discrete intensity maxima are observed in bulk diffraction.

That is, with a given interlayer spacing, *d*, diffraction will only occur at a series of discrete energies when the Bragg condition is satisfied. However, as diffraction in LEED occurs largely from the outermost five or six layers, periodicity in the *z*-direction is far from perfect. In fact, the oscillatory relaxation in the outermost layers (see Fig. 1.20) further reduces the periodicity sensed. This gives rise to a relaxation of the condition required for Bragg reflection and a broadening of the Bragg reflections. Figure 2.25 illustrates the reciprocal lattice rods normal to a surface in a particular surface direction (no periodicity in the *z*-direction for a two-dimensional lattice and hence all values of *n* allowed). Since the intersection of the Ewald sphere (dotted circles of radii $|\boldsymbol{k}_0|$) with a reciprocal lattice rod will occur at *all* values of $|\boldsymbol{k}_0| > 2\pi/a$, a particular diffracted beam should be equally intense at *all* electron beam energies. In contrast, for X-ray diffraction, in which X-rays penetrate and scatter off all layers (bulk diffraction), the perfect periodicity in the direction normal to the surface is sensed and the reciprocal lattice rods become points. In this case, as the beam energy is increased, the beam intensity will be zero at most energies until the exact electron wavelength for bulk diffraction is met (three Bragg conditions satisfied simultaneously for *x*-, *y*-, and *z*-directions, or, more straightforwardly, when the Ewald sphere intersects a reciprocal lattice point).

For a real surface, an intermediate situation between a two-dimensional layer and a three-dimensional crystal arises. Consequently, although the diffraction intensity from a truly two dimensional layer should not vary with

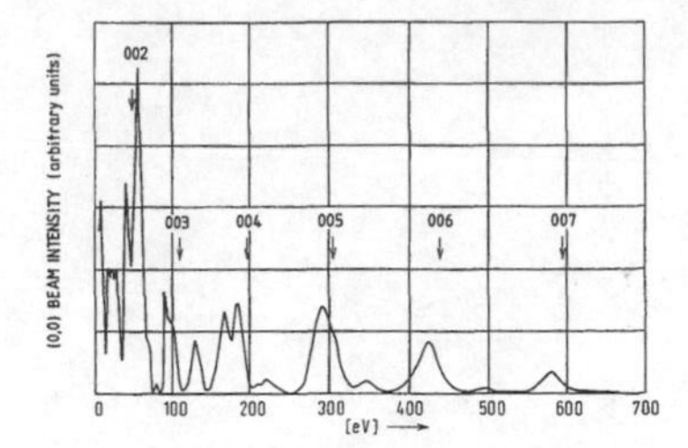

Fig. 2.26 $I(V)$ spectrum from the (0,0) beam of Ni(100). The expected positions of the bulk Bragg diffraction peaks are indicated by the arrows. Ref 4.

energy, because scattering from the outermost five or six surface layers are involved, strong intensity maxima occur when all three Bragg conditions are satisfied.

Figure 2.26 illustrates an intensity–energy or $I(V)$ spectrum for the (0,0) reflection from a Ni(100) single crystal close to normal incidence. The Ni(100) surface provides a surface termination without significant relaxation, hence an almost constant interlayer spacing results. Marked on the spectrum are the expected 'Bragg positions', based on eqn 2.30.

The experimentally measured locations of the 'Bragg' peaks are somewhat shifted from the positions predicted on the basis of the model outlined above and it should be noted that significant extra structure is present within the experimentally measured spectra. The shifting of the Bragg-like peaks can be partially explained by noting that the diffraction effects giving rise to the peaks observed within $I\ (V)$ spectra occur *within* the solid. As an electron enters a solid it experiences a drop in potential energy owing to the electrostatic attraction of the ion cores. In LEED terminology it is said to experience the 'inner potential' (V_r). Inner potential magnitudes vary from metal to metal but usually are equal to ~10 eV. Since energy must be conserved in diffraction, as the electron enters the solid, a drop in potential energy by V_r must be offset by an increase in the electron kinetic energy. Since diffraction occurs within the solid with an increased kinetic energy (E') relative to that in vacuum (E), the energies at which constructive interference should occur relative to the vacuum level will appear to be lowered by a value equal in magnitude to that of the inner potential (V_r).

$$E' = (E + V_r) = \frac{150.6n^2}{4d^2} \tag{2.31}$$

However, this still fails to explain quantitatively the exact positions of the Bragg peaks and offers no explanation as to the origin of the extra features measured in the spectra.

The deviation is actually a result of the fact that electrons interact very strongly with a solid surface and there is a high probability of 'multiple scattering', i.e. an electron can scatter elastically several times before leaving the solid. Thus, it is not correct to set the path length difference simply to $2d$.

In addition to this 'multiple scattering' effect, the strong Coulomb potential of the ion cores of the scattering surface atoms leads to a 'phase shift' of the electrons as they scatter from an ion core. As an electron approaches the ion core, its potential energy is lowered because of its strong attraction with the protons of the nucleus. Conservation of energy requires an increase in kinetic energy and a corresponding decrease in the de Broglie wavelength. This leads to a shift in the phase of the electron wave (Δ) relative to that expected if no encounter with a nucleus was present, as shown in Fig. 2.27. Consequently, the interference between electrons is dependent not only on the path length difference but also on the strength of scattering by ion cores via the phase shift. Thus, the measured beam intensity, as a function of kinetic energy, will depend on the relative phases of interfering waves. This, in turn, is related to the atomic scattering properties of surface atoms *and* the surface geometry.

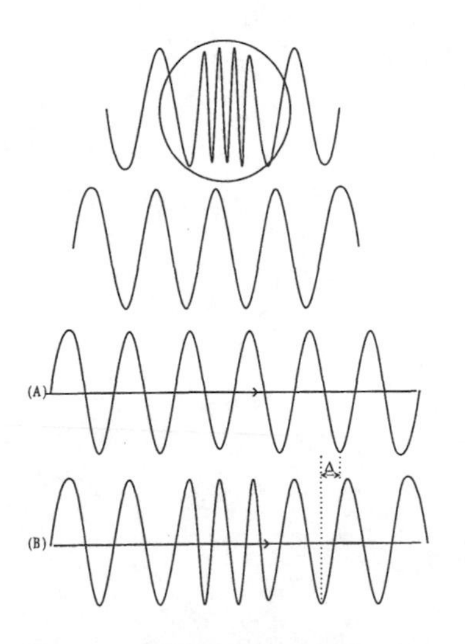

Fig. 2.27 Top: Change in kinetic energy of electron (decrease in λ) as it experiences the 'inner potential' of the solid. The unperturbed wave is also shown. A+B. Phase shift in electron wave resulting from scattering off ion cores (Δ).

To disentangle the atomic scattering phase shift and account properly for the complex 'multiple scattering' is a non-trivial task and precludes any direct route from measured $I\ (V)$ spectra to the surface geometric structure. Instead, a trial and error process is adopted. The flow diagram in Fig. 2.28 illustrates the iterative process used to 'solve' surface crystal structures. The process relies on location of the 'best fit' between experimental spectra and those predicted for a postulated structure. This is generally done using a reliability or '*R*' factor analysis, a method of automatically judging the level of agreement between theory and experiment. Evidently, while LEED has been and remains the premier technique for elucidating surface crystallography, the words of J. J. Berzelius should always be kept in mind:

> To find the truth is a matter of luck, the full value of which is only realised when we can prove what we have found is true. Unfortunately, the certainty of our knowledge is such that all we can do is follow along the lines of greatest probability.

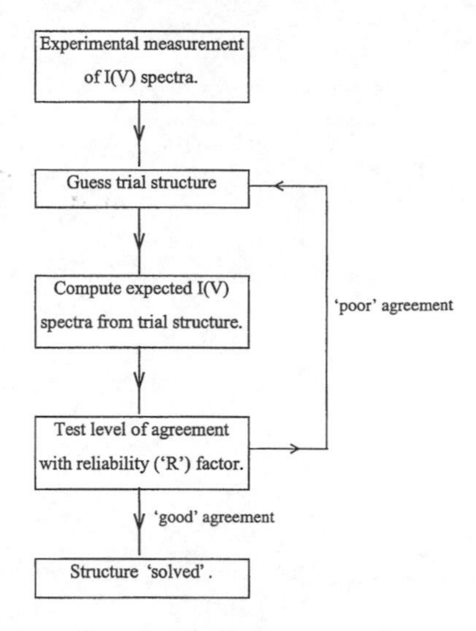

Fig. 2.28 Flow diagram showing the iterative process used to elucidate surface crystallography from LEED $I(V)$ data.

2.4 Scanning probe microscopies

The scanning tunnelling microscope (STM) was invented by Gert Binnig and Heinreich Röhrer at IBM, Switzerland, in 1982, for which they received the Nobel Prize for Physics in 1986. The principle of STM is simple. An atomically sharp tip is brought within a few nanometres of a conducting surface and a small potential difference is applied between the tip and the sample. If the tip is biased positively relative to the sample an energetic incentive is provided for electrons from the sample to flow to the tip, where their potential energy will be lowered. That such a current should flow across 'free space' is surprising since, classically, the electrons in the sample are bound within the solid and a minimum amount of energy, equal to the sample work function ϕ (typically several eV) must be supplied. At room temperature, the average thermal energy available is only of the order of tens of meV. However, at small tip–surface separations, instead of surmounting the activation energy barrier for electron transfer (the work function), electrons are able to pass through the small vacuum gap by a process known as 'electron tunnelling'. Electrons are thus able to flow between the tip and the surface, producing a small, yet measurable current. The magnitude of this current is exponentially dependent on the tip–surface separation. The larger the distance between the tip and the surface, the smaller the current. Therefore, by measuring the magnitude of the tunnelling current as the tip is moved across the surface, a topographic image of the surface is obtained. In favourable circumstances, **atomic resolution** can be achieved.

Figure 2.29 illustrates an energy level diagram for a tip close to a conducting surface in which the tip and sample material are identical (same value of ϕ). The full line shows a wavefunction of an electron in the highest occupied energy level of the sample. If an infinite potential energy barrier was present at the surface, the amplitude of the electron wave would be zero outside the metal and, hence, the classical result would pertain, i.e. an electron could not escape from the solid. However, for a finite barrier height such as in a real metal, the wavefunction penetrates beyond the sample (see also Section 2.5) in such a way that the electron density gradually drops to zero at distances of nanometres outside the surface. Thus, if a second metal is brought to within a couple of nanometres of the surface, there is a finite probability that an electron can 'tunnel', i.e. 'hop' from the sample to the tip where it will lower its energy owing to the positive potential that has been applied to the tip. The positive potential shifts the electronic energy levels on the tip to lower

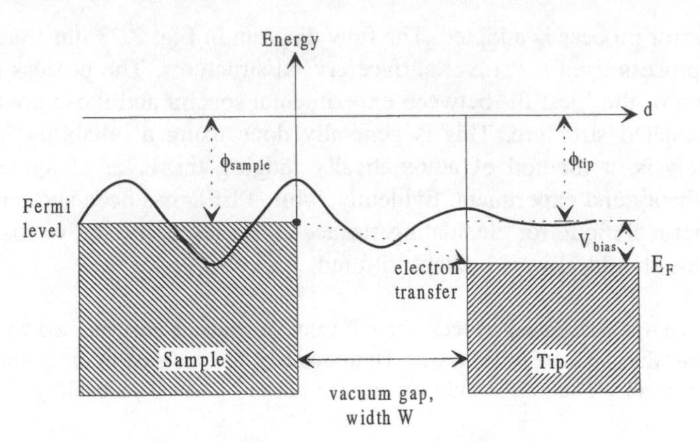

Fig. 2.29 Energy level diagram showing electron tunnelling between tip and sample in STM.

potential energy and thus facilitates electron transfer into unoccupied states of lower energy. Of course, there is no reason why one could not bias the tip negatively and engender electron transfer from tip to sample!

The tunnelling current (I) depends exponentially on the sample to tip gap (W) and the sample work function (ϕ)

$$I(W) = C\exp(-W\sqrt{\phi}) \tag{2.32}$$

where C is a constant. Thus, if a tip is scanned at constant height above a surface the tunnelling current will increase in areas where protrusions exist because of a lowering in the gap distance.

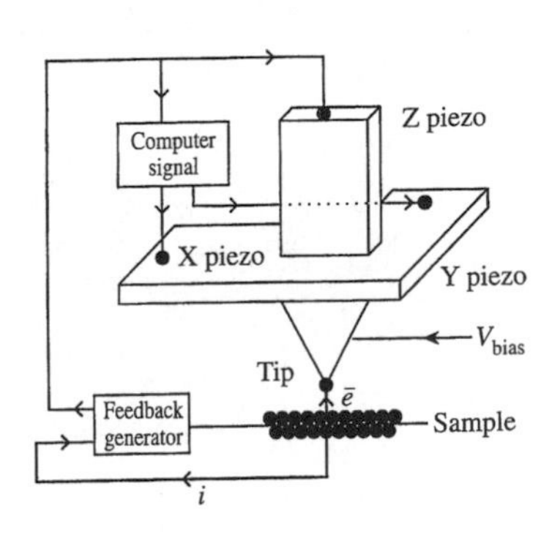

Fig. 2.30 Experimental apparatus used in STM.

Figure 2.30 illustrates schematically the components of a scanning tunnelling microscope. Surprisingly, tips containing a single atom at its apex such as the one shown in Fig. 2.30, may be formed by simply cutting a Pt/Ir wire with a sharp pair of scissors. Alternatively, a tungsten wire may be electrochemically etched in an aqueous solution of sodium hydroxide. The tip quality is tested by its ability to produce good atomically resolved images of standard calibrant materials, such as the basal plane of graphite. The tip is then mounted in the STM head on a piezoelectric tube scanner. A piezoelectric material has the useful property that when a voltage is applied across it, the material expands or contracts. Typically, the piezoelectric will expand/contract by ~1 Å per millivolt, yielding ultra-high precision in tip placement relative to the surface. The piezoelectric tube may be built in three parts such that application of voltages to each element individually gives the possibility of simultaneous movement in the x-, y-, and z-directions with a high degree of accuracy. Clearly, any vibrations transmitted to the tip from the surroundings will lead to a possible collision between the tip and the surface, hence system damping must be employed to prevent acoustic vibrations within the building reaching the sample or tip. This may be achieved simply by placing a series of elastomer sheets underneath the STM and between the tip and its housing.

Two modes of scanning may be employed. In the 'constant height' mode, the tip is scanned in the xy-plane of the surface whilst remaining stationary in the z-direction. This results in variations in the tunnelling current associated with changes in W (owing to surface protrusions). Hence, an image is produced consisting of tunnelling current variations as a function of position in the surface plane, which reflects the surface topography. In contrast, for the

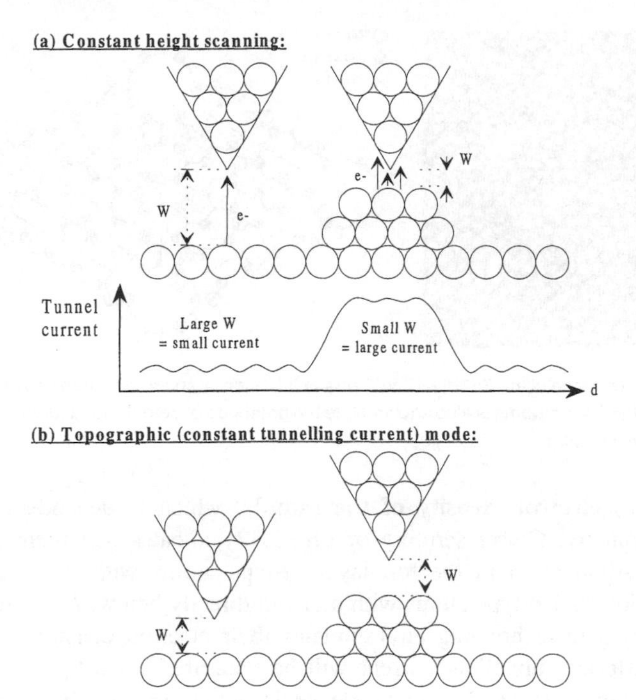

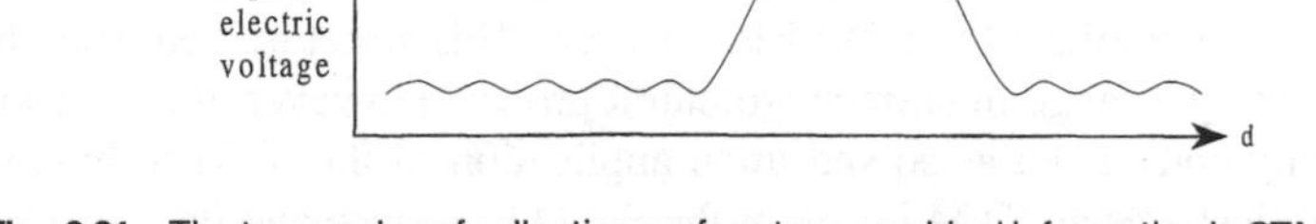

Fig. 2.31 The two modes of collecting surface topographical information in STM: (a) constant height, (b) constant current.

'constant current' mode, the value of W is *fixed* by movement of the tip in the z-direction, whilst scanning in the x y-plane. Thus a plot of the z-piezo electric voltage *versus* lateral position will also yield a topographic image of the surface. The constant height mode is favoured when examining atomically flat surfaces, since rapid scanning is possible and the tip does not have to move up and down. However, for rough surfaces, the constant current mode is preferred since it avoids tip–surface collisions that would otherwise result in a blunt tip incapable of atomic (or any other) type of imaging. Figure 2.31 illustrates the principles of both types of scanning. Typically, the STM has a resolution of about 1 Å in the plane of the surface and < 0.1 Å perpendicular to the surface, and operates with tunnelling currents ranging from 100 pA to 30 nA.

A benchmark for the quality of the tip and the instrumental stability exhibited by the STM operating in an atmospheric environment has become the imaging of the basal plane of highly oriented pyrolytic graphite (HOPG). Such a surface offers large (~ 1000 × 1000 Å) areas of defect-free regions separated by atomic steps of height < 10 Å. Figure 2.32 illustrates a schematic of the structure of graphite and the basal plane of HOPG. It may be anticipated that the repeat distance sensed by the STM will be the 1.42 Å carbon–carbon nearest neighbour separation. However, thinking in this way would disregard the principle of operation of STM. The STM current depends on the **lateral**

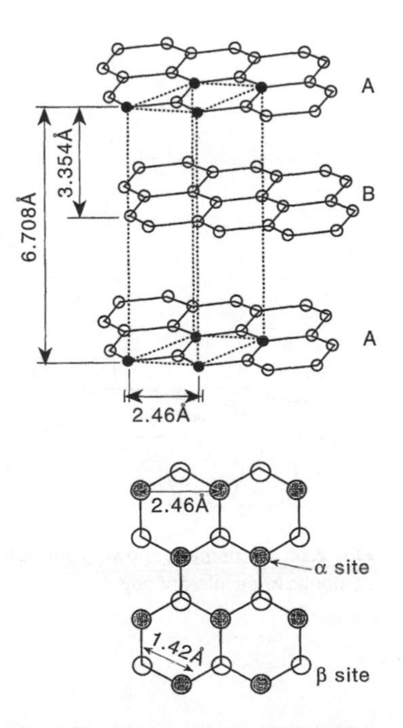

Fig. 2.32 The structure of HOPG. Note that the repeat distance sensed in STM is that between the shaded atoms (2.46 Å), not the interatomic carbon–carbon distance (1.42 Å).

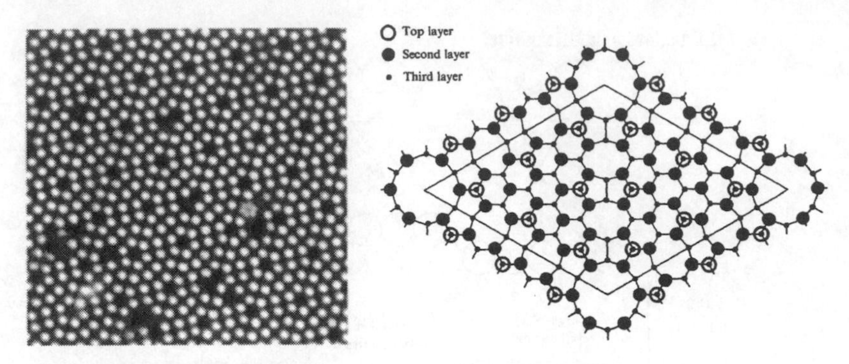

Fig. 2.33 STM image of the Si(111) – (7 × 7) phase [5]. The real space structure showing first, second, and third layer atoms is also shown. Note the presence of defects such as vacancies within the STM image.

variation of electron density of the sample, which is dependent on the surface geometry. Closer scrutiny of Fig. 2.32 indicates that there are two types of carbon atom in the top layer, α-type atoms with no neighbours directly below and β-type atoms with an atom directly below. As these atoms have differing local bonding environments their electron density will vary, hence a differing tunnelling current will be measured for a tip scanning at constant height over α and β atoms. Hence the actual periodic distance sensed in STM is 2.46 Å. One of the earliest and most striking successes of STM was the solution of the structure of a semiconductor surface. The atomically clean Si (111) surface exhibits a (7 × 7) LEED pattern. This structure had been the focus of study by a range of surface structural probes. However, the structure was only truly considered as solved upon application of the STM technique. Figure 2.33 illustrates an STM image of the Si (111) surface and its real space structure. The seven-fold periodicity arises from 'holes' of missing first and second layer Si atoms.

A great advantage of STM over less direct diffraction-based structural probes such as LEED is that it can be utilized not only in vacuum environments but also in air. The STM can also operate in a liquid environment and has recently been utilized to examine electrochemical processes associated with metal deposition from solution on to electrode surfaces. As STM images can be collected in times as short as 10 seconds, images serve as 'snapshots in time' of kinetic processes occurring on surfaces over this time-scale. While STM is the ultimate surface microscopy, it suffers from one major drawback, which arises from the fact that its operation requires a current flow between tip to sample or *vice versa*, i.e. the substrate needs to be a conductor or semiconductor. For insulating materials, STM cannot be used. This problem has been circumvented by a variant on the STM technique called atomic force microscopy (AFM). Figure 2.34 illustrates the principle of operation of AFM. A tip, typically constructed from silicon nitride, with diameter between 1 and 20 nm, is mounted on a cantilever of force constant between approximately 0.001 and 0.2 Nm^{-1}. It is important that the cantilever is insensitive to vibrations and acoustic noise from the laboratory. Hence it is chosen to have a natural resonant vibration frequency as far removed from those experienced in buildings as possible. As the resonant frequency (ν) is given by the classical result

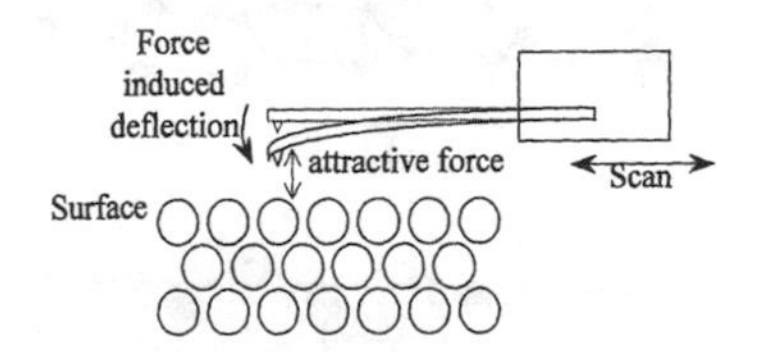

Fig. 2.34 Schematic showing principle of atomic force microscopy.

$$\nu = \frac{1}{2\pi}\sqrt{\frac{K}{m}} \tag{2.33}$$

where K is the cantilever force constant and m its mass. In order to obtain the highest resonant frequency, cantilevers are made with very low masses, typically around 1 μg, with low force constants ($K = 0.004$ Nm^{-1} yields a resonant frequency of 8 kHz).

The AFM tip, when brought into contact with a surface experiences a very small force (of the order of nanoNewtons) as a result of interaction with the surface atoms. In this mode, known as the 'contact mode', the tip is scanned at a tip–sample separation corresponding to a chemical bonding length of the tip/sample combination. This leads to the cantilever being either attracted or repelled as it is scanned across the surface. The repulsive force at very small tip–surface separations originates from both nuclear repulsion and Pauli repulsion as closed electronic shells of surface and tip atoms are forced to interpenetrate. The attractive force arises as a result of the decrease in potential energy of the system caused by chemical bonding through electron overlap between tip/surface atoms. The deflection of the cantilever may be sampled (force imaging) or kept constant using a feedback loop to keep the force at a pre-set value (constant force topography).

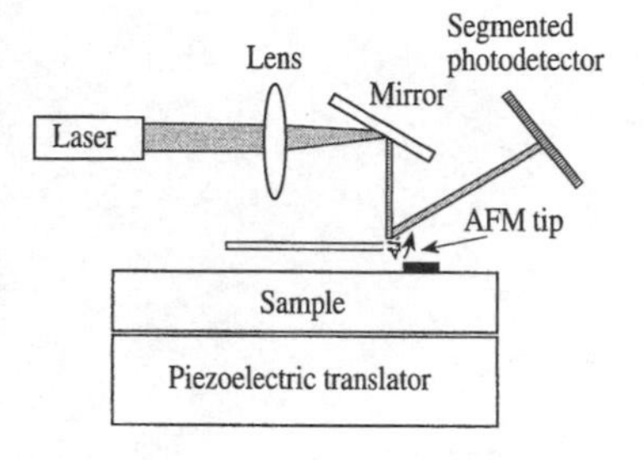

Fig. 2.35 Experimental configuration used to detect force acting on AFM tip.

The deflection of the cantilever may be monitored by an optical technique such as that shown in Fig. 2.35 in which a laser beam is reflected from the back of the cantilever on to a segmented photodetector. As the sample is scanned across the tip, the deflections in the beam arise from changes in local surface topography or 'stiffness', i.e. a soft area of the sample that is easily deformed by the tip will appear as a cavity or pit in the AFM image.

A second mode of scanning, known as the 'non-contact mode' may also be used. This is particularly important for delicate samples which may be damaged by imaging in the 'contact' mode. In this case the tip is not 'in contact' with the surface since no electron cloud overlap occurs. The forces are electrostatic in origin (or magnetic if the tip/sample combination is magnetic) and are even smaller than in the contact mode. In order to enhance the sensitivity of the technique the tip is forced to vibrate close to its resonance frequency and hence this method of measuring is often referred to as the 'tapping mode'. Variations in the sample–tip forces will alter the resonant frequency of the tip, and this frequency shift is used to give a measure of the magnitude of the forces in action. Obviously, for a surface protrusion, the forces acting on the tip will be large so, once again, topographic images of surface force *versus* lateral position on the surface are possible even with non-conducting samples.

Clearly AFM has a wide range of possible applications, ranging from the study of the atomic structure of insulating solids like mica and sodium chloride to polymer coatings on surfaces. Great excitement in the field of biology has been generated with the invention of AFM because of its potential use in imaging biological systems in physiological environments. There are, of course, problems created by the tip–surface interaction and worries concerning the possibility of tip-induced movement of molecules across the sample surface. However, despite such problems, many important observations have been made.

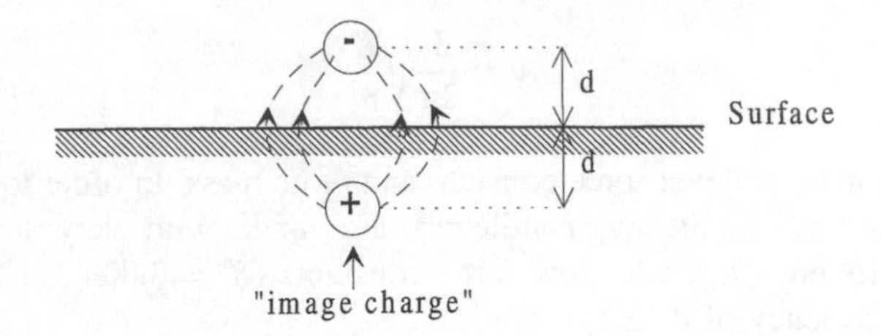

Fig. 2.36 The inducement of an 'image' charge in a surface of opposite polarity to a test charge distance *d* from the surface.

2.5 Work function changes

The work function (ϕ) is defined as the minimum energy required to remove an electron from a solid, i.e. to take an electron from the highest occupied level (the 'Fermi level'), to a sufficient distance outside the surface such that it no longer feels the effect of the long-range Coulomb interactions with its 'image charge', i.e. the positive hole localized at the metal surface, created as a result of removal of the electron (Fig. 2.36; see also figure 1.23).

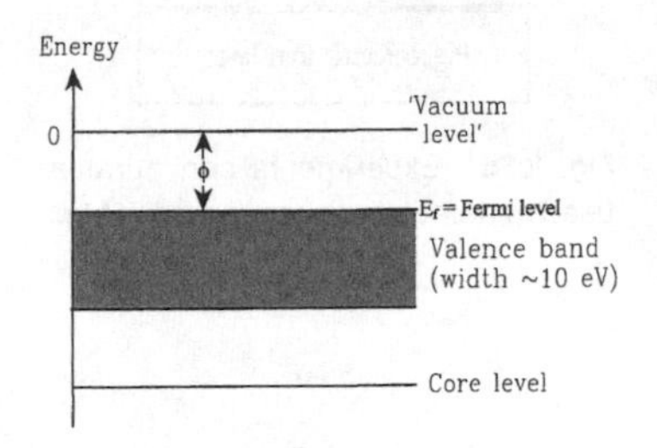

Fig. 2.37 Schematic of energy levels in a solid.

Figure 2.37 shows schematically energy levels in a metallic solid. As the Fermi energy of a solid is a **bulk property**, associated with the electrostatic attraction between atomic nuclei and the valence electrons, at first glance it is not apparent how ϕ yields surface information. However, measurements of ϕ have shown that atomically clean single crystal surfaces of differing geometric structure exhibit different work functions. This is due to the fact that a surface does not present an infinite potential energy barrier to the electrons within a solid. Although the electrons are bound in the solid, the electron wavefunctions themselves may have a non-zero amplitude 'just outside' (for practical purposes within 10 Å) of the surface. Electron wavefunctions are exponentially damped as they penetrate outside the surface and give rise to 'electron overspill'. To preserve overall electrical neutrality, the excess negative charge arising from electron 'overspill' into the vacuum is balanced by a corresponding excess positive charge at the solid surface. Hence, a dipolar layer is formed (Fig. 2.38).

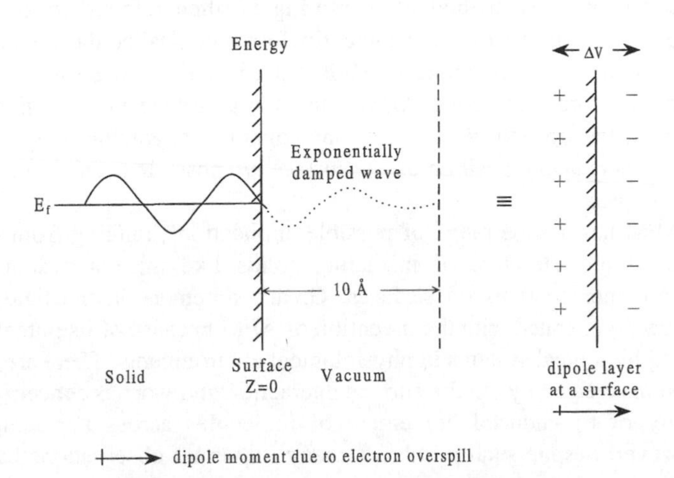

Fig. 2.38 Dipolar layer induced at solid–vacuum interface owing to 'overspill' of electron charge into vacuum (dotted line). The full line indicates an electron wave function in the solid.

The work function can be considered as consisting of two contributions, one associated with the bulk electronic properties of the solid and the second with the surface dipole layer described above. The greater the 'overspill', the larger will be the surface dipole. Since the extent of electron overspill is a function of surface geometry (there are more atoms per unit area in a closed packed surface and hence more electrons available for overspill), the work function is a **surface property**.

For example, for those low index fcc metals, where no surface reconstructions occur

$$\phi_{111} > \phi_{100} > \phi_{110}$$

e.g. $Cu(111){:}\phi = 4.94$ eV, $Cu(100){:}\phi = 4.59$ eV, $Cu(110){:}\phi = 4.48$ eV.

Adsorption may also induce changes in the work function associated with modifications of the surface dipolar layer, particularly if significant charge transfer occurs between the adsorbate and surface. Thus, measurements of the work function *change*, $\Delta\phi$ yield critical information on the degree of charge reorganization upon adsorption ($\Delta\phi = \phi_{\text{adsorbate covered}} - \phi_{\text{clean}}$).

Because the dipolar layer may be modelled as a parallel plate capacitor, the **Helmholz equation** can be used to relate a change in work function ($\Delta\phi$) to the dipole moment, μ, of the adsorbate and its image charge.

$$\Delta V = n\mu/\varepsilon_0 \tag{2.34}$$

where ε_0 = permittivity of free space, = 8.85×10^{-12} C V^{-1} m^{-1}; n = surface density of the adsorbate (m^{-2}); and μ = dipole moment (Cm). ΔV is called the change in surface potential. ΔV may be converted to a work function change (in Joules) by multiplying ΔV by the electron charge, e. Hence if, for example, a surface coverage of 5×10^{14} cm^{-2} atoms of chemisorbed gas on Cu(111) gave rise to a change in work function of 0.4 eV, one could estimate the surface dipole using eqn 2.34

N.B. need to convert cm^{-2} to m^{-2}

1 Debye (D) = 3.3356×10^{-30} Cm

$$\Delta V = \Delta\phi/e = 0.4\ \text{V} = \frac{5 \times 10^{14} \times 10^{4} \times \mu}{8.854 \times 10^{-12}}$$

$$\therefore \mu = \frac{0.4 \times 8.854 \times 10^{-12}}{5 \times 10^{14} \times 10^{4}} = 7.08 \times 10^{-31}\ \text{Cm}$$

$$= \frac{7.08 \times 10^{-31}}{3.336 \times 10^{-30}} D$$

$$= 0.21\ D$$

From LEED, the distance between the plane of chemisorbed molecules and their image charge plane in the metal is of the order of 3 Å. Therefore, since the dipole moment is given by

$$\mu = Q \times d \tag{3.35}$$

Where d = charge separation and Q = partial charge on chemisorbed atoms, one may estimate Q as being equal to

$$\mu/d = 7.08 \times 10^{-31}\ \text{Cm}/3 \times 10^{-10}\ \text{m} = 2.36 \times 10^{-21}\ \text{C}$$

Since $e = 1.6 \times 10^{-19}$ C, this means that a charge of $\frac{2.36\times10^{-21}}{1.6\times10^{-19}} = 0.015$ electrons is estimated to reside on the chemisorbed molecule. Furthermore, the sign of ΔV often reflects the **net direction** of charge transfer. Two extreme cases may be used to highlight this point.

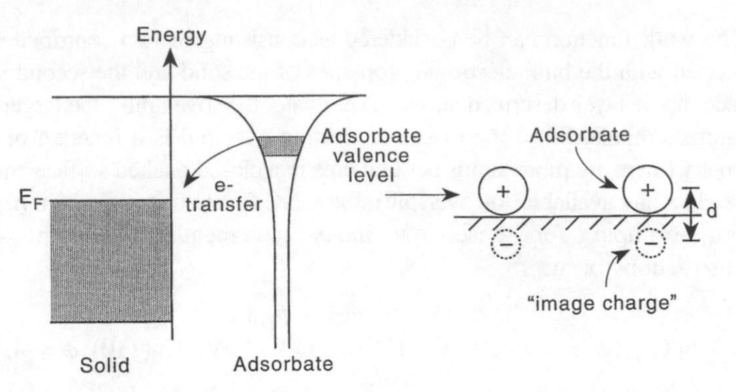

Fig. 2.39 Surface dipole induced by the adsorption of electropositive adsorbates on a metal surface.

Electropositive adsorbates

Alkali metals exhibit low first ionization energies and, when they adsorb on to a solid surface, they tend to transfer electron charge from their outer valence shell to the substrate. This process of electron transfer does not go to completion but ceases when the Fermi energy of the electrons on the substrate is equal to the highest occupied electron state in the broadened valence level of the adsorbate.

Broadening occurs through increasing overlap of the adsorbate and substrate wavefunctions.

Figure 2.39 shows schematically the adsorption of electropositive elements. Because the adsorbate has transferred electron charge to the substrate, a dipolar layer is formed in which a net positive charge now resides on the adsorbate, inducing an equal but opposite image charge at an equivalent distance below the surface plane. Note that the dipolar layer is in the *opposite* direction to the dipolar layer at the clean surface. Hence, a *lowering* in the work function is expected since the *net* surface dipole has been reduced.

Electronegative adsorbates

Electronegative adsorbates usually possess an unfilled affinity level, again broadened into a resonance that is situated largely or entirely ***below*** the highest occupied state of the substrate (Fig. 2.40). In this case, charge transfer occurs in the opposite sense to alkali metals, i.e. from substrate to adsorbate.

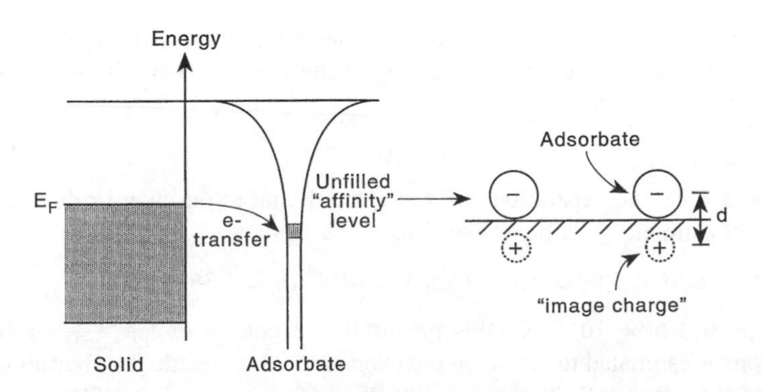

Fig. 2.40 Surface dipole induced by adsorption of electronegative adsorbate on metal surface. Note that induced dipole is now in opposite direction to that in Fig. 2.39 but is the same as in Fig. 2.38.

Consequently, a dipolar layer is formed with a negative charge outermost, i.e. in the same direction as the surface dipole of the clean surface. Hence, the *net*, surface dipole is expected to increase upon adsorption of electronegative elements leading to a **work function increase**. There are a number of methods used to measure changes in ϕ. However, one of the most commonly used techniques involves the measurement of an ultraviolet photoemission spectrum (UPS), discussion of which will be postponed until Section 2.6.

Figure 2.41 illustrates the change in ϕ observed as a function of alkali atom coverage. At low coverage, a linear and rapid decrease in ϕ occurs, which gradually levels off and attains a minimum at a coverage corresponding to approximately half the saturated monolayer coverage. Finally, a slow rise is observed to a value that saturates upon completion of the alkali monolayer, at which point no further changes occur as subsequent alkali layers are deposited. These results have been interpreted as indicative of an insulator–metal transition. The rapid linear decrease is due to large alkali–surface charge transfer, yielding partially positively charged alkali adsorbates. As the surface coverage increases, the positively charged adsorbates are forced closer together, leading to increased repulsive lateral electrostatic interactions. This increasing repulsive interaction leads to 'depolarization', i.e. reduction in the degree of charge transfer, and to a minimum in the curve. Finally, as the coverage approaches monolayer saturation, the adsorbate becomes fully 'metallic', i.e. the outer valence electron charge originally lost through charge transfer to the substrate is returned to the adsorbate, allowing alkali–alkali metallic bonds to be formed. Thus, the measured work function attains a value close to that of the bulk alkali metal.

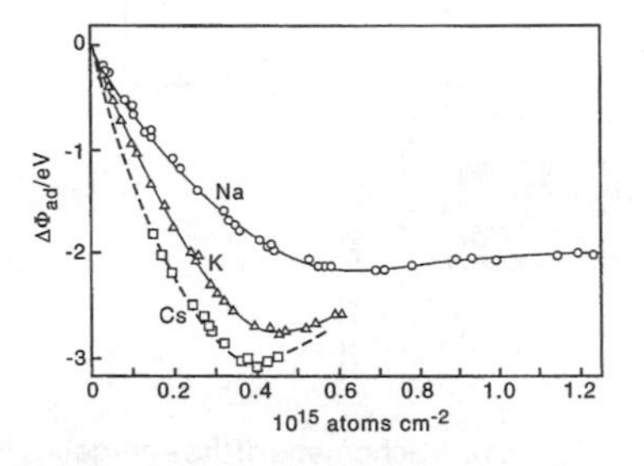

Fig. 2.41 Plot of change in work function as a function of coverage for Na, K, and Cs. Adapted from ref. 6.

The models outlined above to explain $\Delta\phi$ are clearly oversimplistic. For example, an important aspect of work function measurements is that the change in the direction of the surface dipole will depend on whether the adsorbate remains *above* the surface or moves into subsurface sites; i.e. although an electronegative adsorbate would be expected to increase the value of ϕ, if the adsorbate occupied subsurface sites, the adsorbate-induced surface dipole would obviously be in the opposite direction leading to a decrease in ϕ. Despite this, work function measurements do offer a unique insight into the electronic charge rearrangement that occurs upon formation of a surface chemical bond. Even greater insight, however, may be obtained from the use of ultraviolet photoemission spectroscopy (UPS).

2.6 Ultraviolet photoemission spectroscopy (UPS)

A wide range of surface properties are controlled by the loosely bound valence electrons of surface atoms and molecules. For example, it is the increase in energy experienced by these valence electrons upon loss of bulk coordination when a surface is created that determines the 'surface energy'. The distribution of charge at the solid/vacuum interface also determines the value of the work function. Finally, the strength of adsorbate–surface bonds and the stability of reactive intermediates in catalysis requires a detailed understanding of the valence energy levels of both adsorbate and substrate. The UPS technique is related to XPS but, whereas XPS is used for elemental identification by study of the strongly bound 'core' electrons, the domain of UPS is the study of the weakly bound valence levels that participate in chemical bond formation.

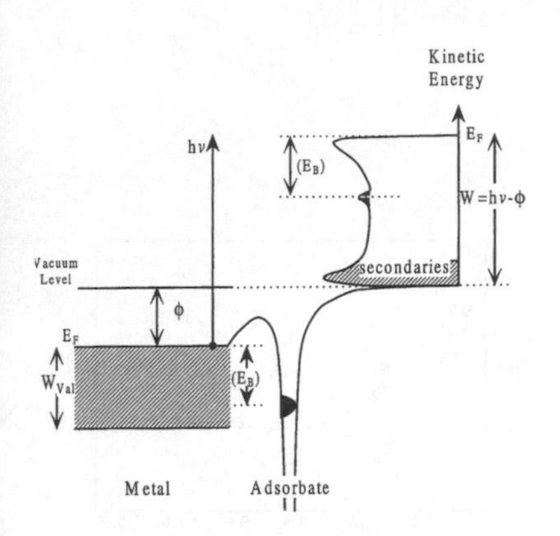

Fig. 2.42 Schematic of the energetics of an ultraviolet photoemission experiment.

Since, in general, a large number of valence levels (originating both from the substrate and the adsorbate) are contained within a rather narrow energy range (typically ~ 10 eV), X-ray photon sources are unsuitable for valence band studies since their inherent energy spread of 1 eV leads to poor resolution of valence peaks. Lower energy UV photons exhibit a much narrower energy width and hence are more useful in studying valence band structure. Low energy UV-photons are produced by striking an electrical discharge inside a low pressure of noble gas. As electronically excited noble gas atoms relax back to their ground state, they emit highly monochromatic UV photons. The most popular sources are helium and neon discharges, whose primary emission lines yield photons of energy 21.22 eV (HeI) and 16.85 eV (NeI). As most valence levels of the substrate and adsorbate are observed between the Fermi level to a binding energy of 10 eV, such sources are sufficiently energetic to excite photoemission from valence levels. Energy analysis of the photoemitted electrons can be performed with an electrostatic analyser as described previously in the section on XPS.

Figure 2.42 illustrates photoemission from the valence band of a solid and from an adsorbate with a single valence level using monochromatic UV photons of energy $h\nu$. The highest kinetic energy electrons are emitted from the Fermi level. Superimposed on the emission from the substrate valence band are electrons photoemitted from the weakly bound valence level of the adsorbate. The binding energies of these levels can be measured by UPS using the Einstein equation in an analogous manner to XPS

At low kinetic energies, an intense broad peak is always observed in UPS owing to emission of 'secondaries'. As mentioned in Chapter 1, secondary electrons arise from directly emitted photoelectrons initially having kinetic energies as described by the Einstein equation, which lose energy through excitation of plasmons and phonons as they pass through the solid, *en route* to the detector. As they have undergone a multiple excitation process, they provide limited surface information and can be 'ignored'.

$$E_{\mathrm{b}} = h\nu - E_{\mathrm{Kin}} - \phi$$

Electronic structure of clean surfaces

The form and shape of emission from the substrate yields information on the substrate electronic structure. As mentioned earlier, the outer valence electrons of a solid will form an 'energy band'. In first row transition metals, for example, the band arises from overlap of the outer 3d and 4s electrons, forming a 'd band' and 'sp band'. As the outer sp electrons are the most diffuse orbitals, they undergo the largest overlap with neighbouring atoms and, hence, exhibit the broadest bands. The more tightly bound d electrons have a less effective overlap and, hence, form a narrow band with a high density of electronic energy levels ('density of states'). In contrast, as the sp band is spread over a larger energy range, it exhibits a low density of states. Figure 2.43 illustrates the situation in which the narrow 'd band' is located in the middle of the sp band. The position of the Fermi level is dependent upon the number of valence electrons. For example, Fig. 2.44 illustrates the electronic structure for an open d band metal such as palladium, in which the d band is partially filled and the Fermi level lies within the d band. This is signalled in UPS by intense photoemission at the Fermi level. In contrast, for Cu, the d band is filled and the Fermi level cuts the sp band, leading to weak emission from the low density of states sp band and an intense d band emission centred at 2–3 eV below the Fermi level from the filled d band. Thus, UPS from the substrate allows qualitative conclusions to be drawn about the substrate electronic structure. For example, high catalytic reactivity is usually

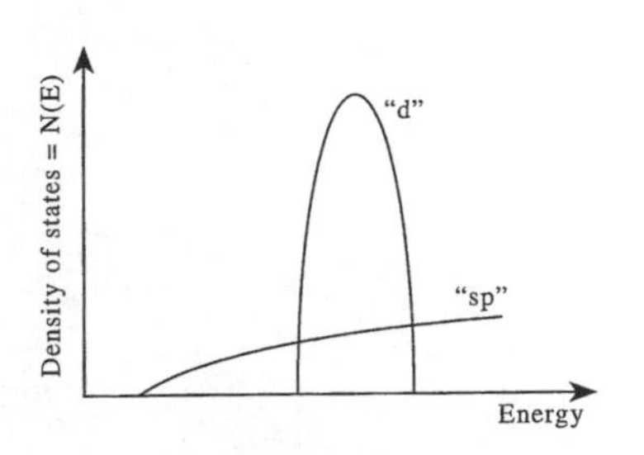

Fig. 2.43 Schematic of the variation in density of states versus binding energy for a transition metal.

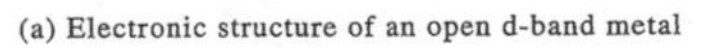

characterized by a high density of electron states at the Fermi level. Thus, for example, metals such as Cu, Ag, and Au, which have filled d bands, are generally rather unreactive and form only relatively weak chemical bonds with adsorbates relative to open d band metals such as Pd or Pt, which are excellent catalysts.

In fact, control over the surface electronic structure (and hence the surface reactivity of a solid) may be achieved by epitaxial growth of one metal upon the surface of another. This facilitates both modification of the electronic properties of the surface through charge transfer to the substrate, and also modification of the lattice constant of the adsorbate to match that of the underlying substrate. A second method of tailoring the electronic properties of surfaces is by alloying. For example, when half a monolayer of palladium is deposited on a Cu(100) surface, formation of a 'surface alloy' results associated with Cu/Pd intermixing in the surface. Such a surface, while containing a large number of palladium atoms, exhibits remarkable chemical properties. For example, while carbon monoxide normally adsorbs strongly on palladium surfaces at room temperature, the alloy surface is found to be incapable of adsorbing CO at 300 K. UPS has provided the answer to this puzzle. The emission from the surface alloy at the Fermi level is weak and very similar to pure Cu. By comparing the UP spectra from pure Cu and the alloy, the Pd d band emission is centred at ~ 1 eV below the Fermi level, indicating a shifting of the Pd d electronic states to higher binding energy in the alloy relative to a pure Pd surface. This means that the Pd d band in the alloy is now full. In essence, a Pd surface has been produced that is 'Cu-like', i.e. possessing a full d band. Such a surface has been shown to have adsorption properties *intermediate* between those of pure Cu and pure Pd. Hence, the opportunity of artificially altering the surface electronic structure allows 'tuning' of the adsorption and catalytic properties of a surface.

(a) Electronic structure of an open d-band metal

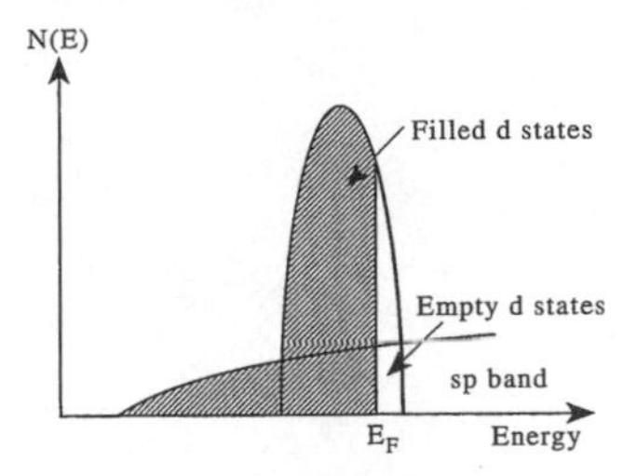

(b) Electronic structure of a filled d-band metal

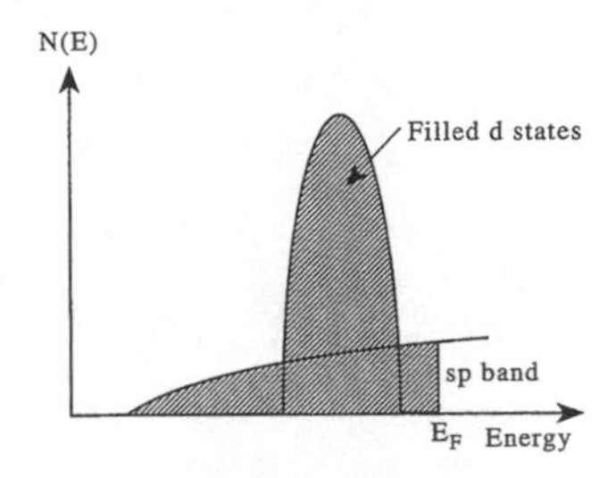

Fig. 2.44 (a) High density of electron states at E_F for metals with open d bands. (b) Low density of electron states at E_F for metals with filled d bands.

UPS as a probe of molecular adsorbates

UPS allows information to be gained on the molecular state of an adsorbate using a technique known as 'fingerprinting'. This relies on a comparison of the UP spectrum from an adsorbate with its corresponding UP spectrum obtained from the gas phase molecule. In such studies, emission from adsorbate molecular orbitals may be isolated by plotting a 'difference spectrum', i.e. the spectrum obtained by subtracting the clean surface UP spectrum from its adsorbate-covered counterpart. Figure 2.45 illustrates UP spectra from methanol adsorbed on Cu(110) and the corresponding gas phase UP spectrum of methanol. The negative deviation in the spectra from chemisorbed methanol around the Fermi level is due to attenuation of emission from the substrate valence band caused by inelastic scattering as electrons pass through the adsorbed methanol layer. Also shown in Fig. 2.45 is the UPS of gas phase methanol. The methanol valence orbitals are labelled according to their symmetry (which will not concern us here). As all molecular orbitals generally suffer an energy shift when adsorbed on a surface, the gas phase spectrum is aligned with an orbital that is not involved in bonding with the surface. In this case, the 5a′ methanol orbital has been used. As can be seen, the spectrum of methanol chemisorbed at 140 K agrees well with the gas phase spectrum indicating that the adsorbate maintains its molecular integrity when adsorbed. The 2a″ and 7a′ peaks, however, have been shifted to higher binding energy

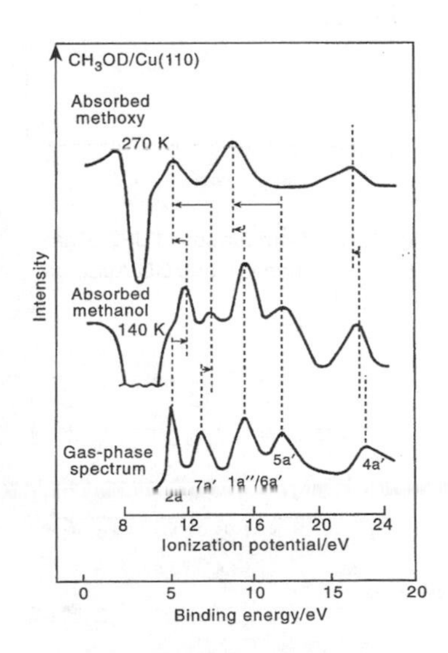

Fig. 2.45 UP spectra of CH_3OD adsorbed on Cu(110) at 140 K and heated subsequently to 270 K. The gas phase UP spectrum of CH_3OD is also shown. See text for details. Adapted from ref. 7.

relative to the other features. These orbitals are lone pairs located on the oxygen atom. Their differential shift can be taken as evidence that they are primarily responsible for the chemisorption bond to the surface. Thus, we may conclude that methanol adsorbs molecularly and is bonded through the oxygen lone pairs to the Cu surface.

This example illustrates the general principle of 'fingerprinting' in which:

(i) the gas phase and adsorbate UP spectra are aligned using a suitable orbital which is non-bonding to the surface; and

(ii) any orbitals that have been substantially shifted may be implicated in bonding to the surface.

The upper spectrum in Fig. 2.45 illustrates the effect of heating the methanol monolayer to 270 K. A radical change in the UP spectrum is seen to have occurred. The four highest occupied orbitals reduce to only two emission peaks. This signals a major change in surface bonding. However, decomposition of methanol to carbon and oxygen can be ruled out since the UPS peaks do not correspond to the energies expected from the 2p orbitals of atomic oxygen and carbon. In fact, the change in UPS has been interpreted as being a result of decomposition of the adsorbed molecular methanol to a methoxy ($-OCH_3$) intermediate bound *via* the oxygen atom to the Cu surface.

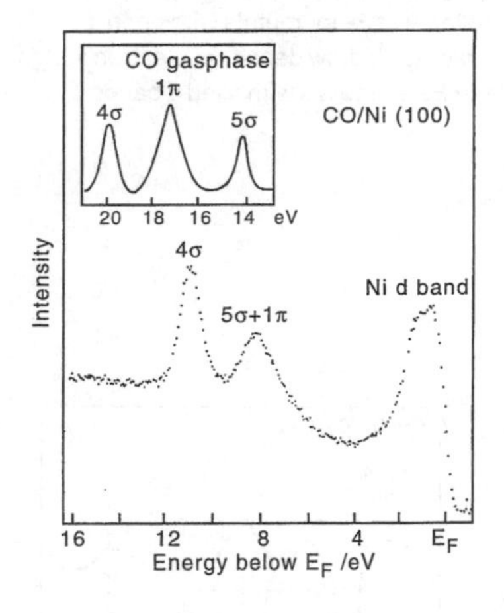

Fig. 2.46 Comparison of UPS of gas phase and chemisorbed CO. Adapted from ref. 8.

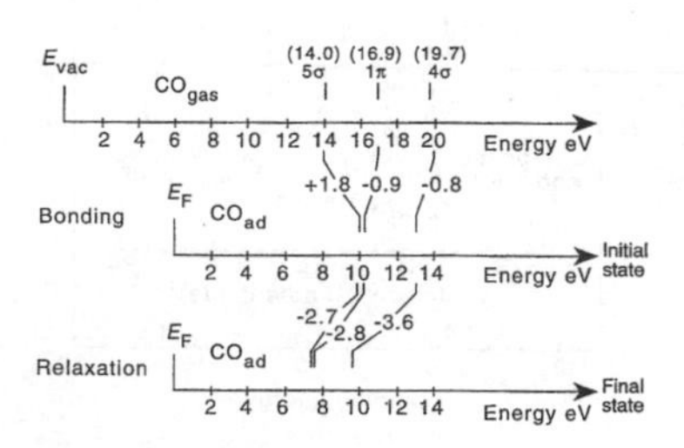

Fig. 2.47 Comparison between shifts induced in CO valence molecular orbital energies due to initial and final state effects. Ref [9].

However, one needs to be careful in *rigidly* assigning relative shifts in UPS peaks from adsorbates purely to changes in bonding (so-called 'initial state effects'). Relaxation of the electronic energies of valence orbitals associated with the presence of a core hole state (created as a result of photoemission) may also play an important role (so-called 'final state effects'; see discussion of inadequacy of Koopman's theory applied to photoemission in Section 2.1). An example of this effect is the adsorption of CO on metal surfaces. Figure 2.46 depicts the main photoemission features from both gas phase and chemisorbed CO molecules, taking into account the fact that the former is referred to the vacuum level, whereas the latter is referred to the Fermi level of the solid. It is evident that the three gas phase peaks, 5σ, 1π, and 4σ, reduce to just two peaks in the chemisorbed phase [8]. It is also seen that, according to the 'fingerprinting' principle as applied to methanol above, it is the 4σ and 1π orbitals that appear to be involved in the bonding of CO to the metal, since these orbitals undergo the largest relative shift compared to the gas phase. In fact, the 5σ and 1π orbitals overlap in the chemisorbed state (hence just two peaks from molecularly adsorbed CO are observed in UPS). However, the bonding in metal carbonyl compounds is rather well understood in terms of the 5σ orbital from the carbon atom donating charge to the metal and back-donation of charge from the metal into the $2\pi^*$ orbital of CO. So why does the 5σ orbital remain essentially unchanged upon bonding to a surface? The answer lies in consideration of both initial (bonding) and final (relaxation) state effects. Figure 2.47 shows the result of a theoretical calculation of the relative contributions of initial and final state effects in the UPS of adsorbed CO. From Fig. 2.47, it is clear that initial state effects do indeed, correspond to a rather large shift in the 5σ orbital of CO to higher binding energies and relatively smaller shifts in the 1π and 4σ peaks to lower binding energies, as expected on the basis of our knowledge of bonding in metal carbonyls.

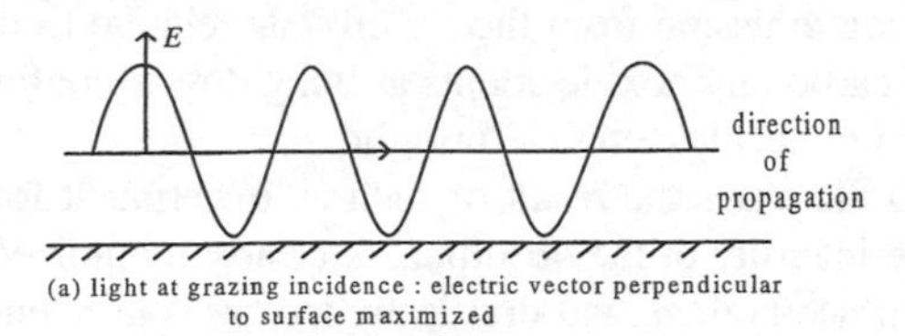

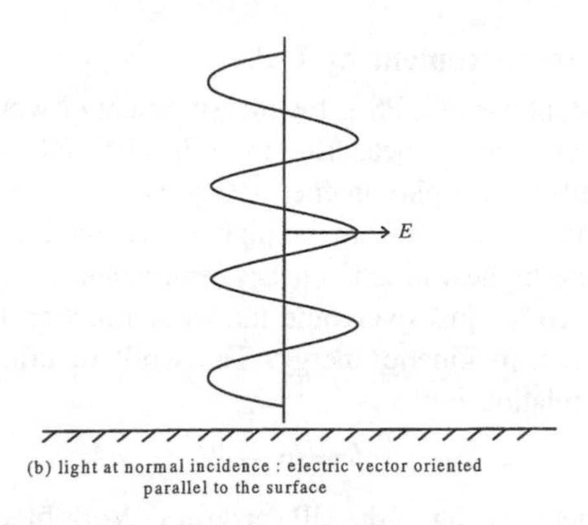

Fig. 2.48 Variation in orientation of electric vector of UV photon as a function of incidence angle.

However, the large relaxation in orbital energies associated with the creation of the core hole shifts the 5σ, 1π and 4σ to lower binding energies. Hence, the *net* shift in the 5σ orbital is small overall, whereas both the 1π and 4σ orbitals have undergone a large shift to lower binding energies compared with their energies in the gas phase.

A further (very useful) application of UPS is in the determination of **molecular orientation**. The orientation of a molecule may be probed by variation of the orientation of the electric vector of the incident light with respect to the intermolecular axis. As light is an electromagnetic wave with an electric field that oscillates in a plane perpendicular to the direction of propagation, the direction of the electric vector with respect to the sample surface may be changed by varying the angle of incidence of the light. Figure 2.48 illustrates this point. For light at normal incidence to the surface, the component of the electric vector parallel to the surface ($\boldsymbol{E}_{\parallel}$) is maximized and the perpendicular component ($\boldsymbol{E}_{\perp}$) is zero. As the angle of incidence is increased, $E_{\perp}$ increases in magnitude. At grazing incidence, a mixture of $\boldsymbol{E}_{\perp}$ and $\boldsymbol{E}_{\parallel}$ exists.

For adsorbates, a simple selection rule applies. Sigma orbitals may be excited with maximum probability when the electric vector of the light is aligned along the internuclear axis, whereas excitation is forbidden when the electric vector is aligned exactly perpendicular to the bond axis. In contrast, π orbitals are excited with maximum probability when the electric vector is perpendicular to the bond axis, whereas excitation is forbidden when the electric vector is aligned along the internuclear axis. To illustrate this point, consider UPS from a CO molecule oriented both in an 'upright' and a 'lying down' configuration for light incident normal to the surface. Use of the selection rule outlined above predicts strong emission from the $4/5\sigma$ orbitals if the molecule adopts a 'lying down' configuration, and zero intensity if the molecule is 'upright' (Fig. 2.49). Conversely, for light incident at grazing

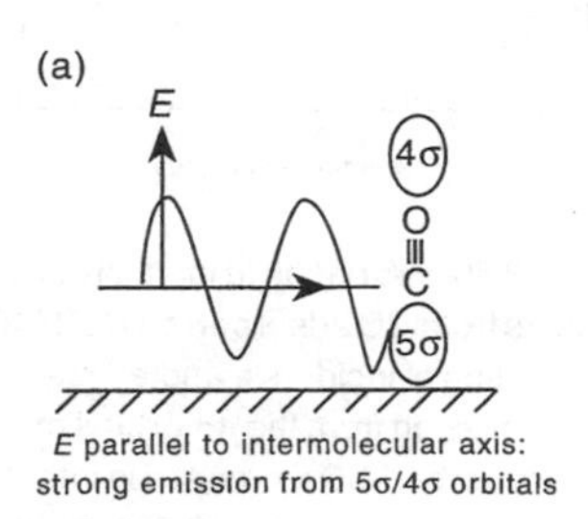

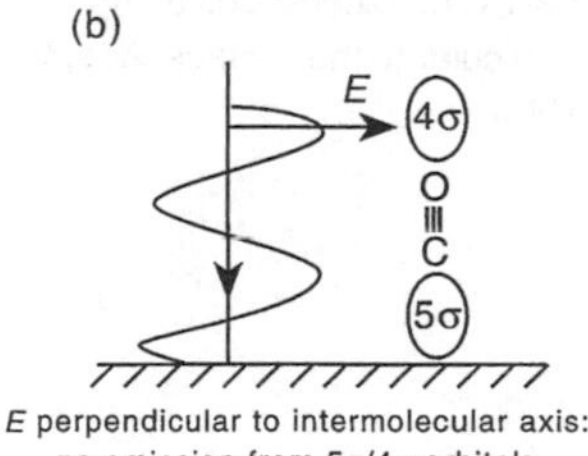

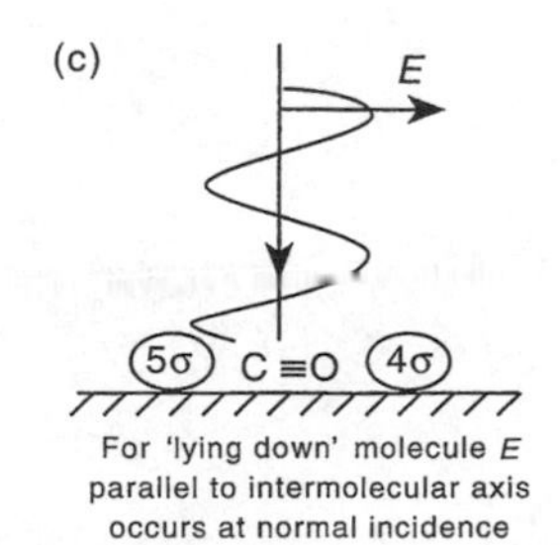

Fig. 2.49 Excitation from σ orbitals as a function of both the orientation of the electric vector and the adsorption geometry of the adsorbate.

incidence, strong emission from the 1π orbitals relative to the $4\sigma/5\sigma$ orbitals, is predicted if carbon monoxide adopts a 'lying down' configuration and weak emission if the molecule remains 'upright'.

Figure 2.50 illustrates the result of such an experiment for CO adsorbed on Co(10$\bar{1}$0). The intensity of the 4σ orbital is clearly minimized for light incident close to the surface normal, and drastically increases in intensity at an angle of incidence of 60°, suggesting that the molecule adopts an upright configuration.

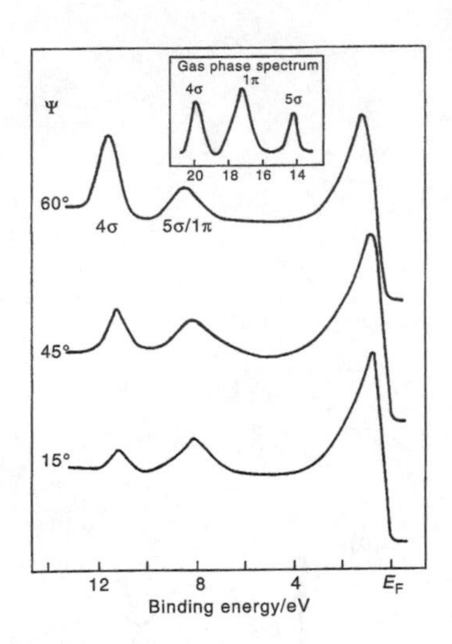

Fig. 2.50 Variation in intensity of UPS peaks from CO adsorbed on Co(10$\bar{1}$0) as a function of incidence angle. Note that the emission from the 4σ orbital of chemisorbed CO increases in intensity as the incidence angle changes from 15 to 60° to the normal. This is consistent with the CO molecular axis being perpendicular to the surface. Adapted from ref. 10.

Work function measurement by UPS

A final and important use of UPS is the measurement of work functions. Work functions are determined by measuring the full width (W) of a UPS spectrum excited with a well-defined photon energy (e.g. HeI, $h\nu = 21.22$ eV), as shown in Fig. 2.51. Electrons emitted from the highest occupied level, the Fermi level (E_F), will yield the highest kinetic energy, equivalent to $h\nu - \phi$. The lowest kinetic energy electrons just overcome the work function barrier and, hence, are 'emitted' with zero kinetic energy. The work function may be derived according to the relationship

$$\phi = h\nu - W \tag{2.36}$$

where W is the energy width of the UP spectrum. Work function changes upon adsorption result in a change in the width (W) of the UP spectrum. An increase in work function leads to a reduction in the UP spectrum width, while, conversely, a decrease in ϕ leads to an increase in the width. The work function change manifests itself in a shift in the low kinetic energy 'secondary tail', as shown in the inset in Fig. 2.51.

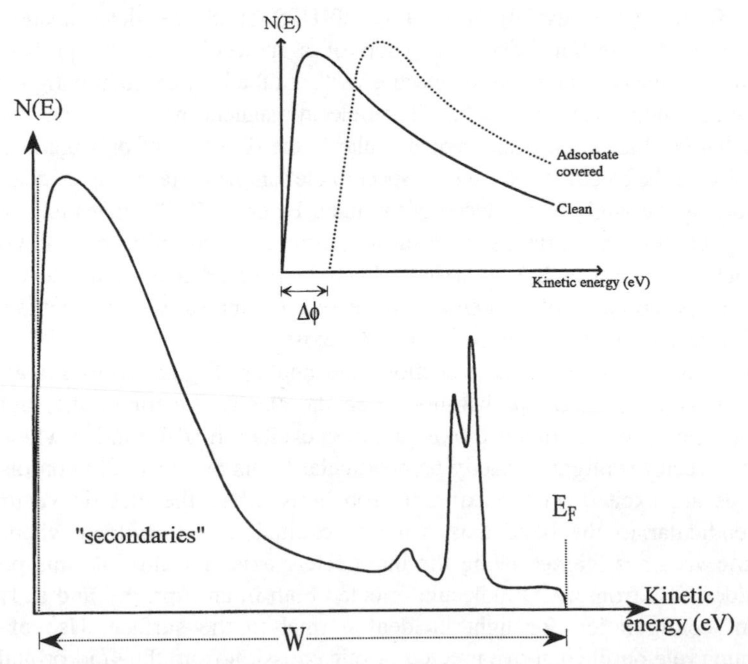

Fig. 2.51 The energy width W of a UP spectrum from a solid may be used to determine the work function. Changes in work function may be obtained from changes in the 'cut-off' of the secondary electron peak (inset).

2.7 Temperature programmed desorption (TPD)

As mentioned in Chapter 1, the desorption of adsorbed atoms and molecules is one of the most fundamental elementary surface kinetic processes and can provide information concerning the strength of the interactions between the surface and the adsorbed species. In temperature programmed desorption (TPD), a temperature ramp is applied to the sample and the rate of desorption is followed by monitoring the amount of adsorbate desorbed into the gas phase as a function of temperature (see Fig. 2.52).

The experimental requirements include the following.

(i) A method of heating the sample in such a way that the heating rate (β) is linear in time (t) and obeys the relationship

$$T(t) = T_0 + \beta t \tag{2.37}$$

where T_0 is the initial sample temperature. The heating should be restricted ideally only to the sample, thus avoiding desorption from other surfaces such as the sample holder. In practice, the most common method of heating is 'resistive heating' in which an electric current is passed through thin support wires spot-welded to the edges of the sample. The wires heat the sample by conduction.

(ii) A method of monitoring the sample temperature. This may be achieved by spot-welding a thermocouple junction to the edge of the sample.

(iii) A detector to monitor the rate of desorption. In most modern surface science laboratories this consists of a **quadrupole mass spectrometer** tuned to the mass of the species being removed from the surface and positioned close to the sample surface in direct line of sight.

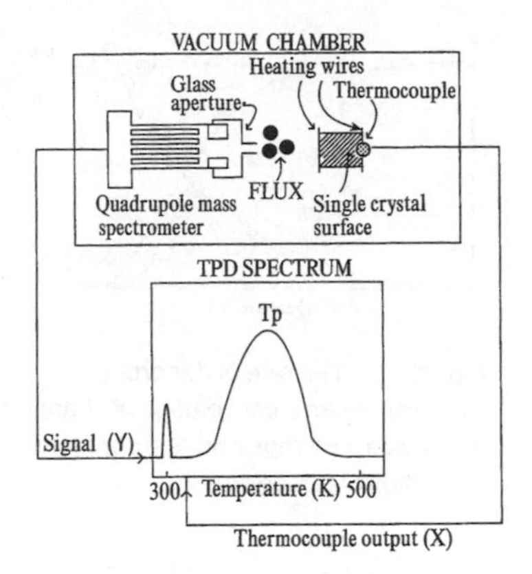

Fig. 2.52 Experimental set-up for performing a TPD experiment. The form of a TPD spectrum is also depicted.

Typical heating rates range between 1 and 100 $\mathrm{Ks^{-1}}$. A glass aperture is often fitted over the ionization region of the mass spectrometer to ensure that only adatoms from the front face of the sample are detected. For mass spectrometers operating *without* a glass aperture, a 'support' peak due to rapid desorption from the heating wires (which heat up much more quickly than the sample) is often seen upon initiation of the temperature ramp. Figure 2.52 also shows the form of an idealized TPD spectrum in a pumped system with a support peak seen upon initiation of the temperature ramp.

As the temperature rises and the thermal energy available becomes sufficient to break surface bonds, desorption is observed. For the simplest case of an adsorbate in which the activation energy for desorption is constant as a function of coverage, a single desorption peak is obtained. Furthermore, as the experiment is performed in a vacuum chamber that is being pumped continually, the temperature at which maximum desorption occurs (T_p), corresponds to the maximum desorption rate. At first sight this may seem unusual since desorption is an activated process with a rate constant (k_d) that obeys an Arrhenius dependency and hence should increase exponentially with temperature

$$k_\mathrm{d} = A\exp\left(\frac{-E_\mathrm{d}}{RT}\right) \tag{2.38}$$

where E_d is the activation energy for desorption and A is a pre-exponential factor. A maximum is observed because, although k_d increases exponentially

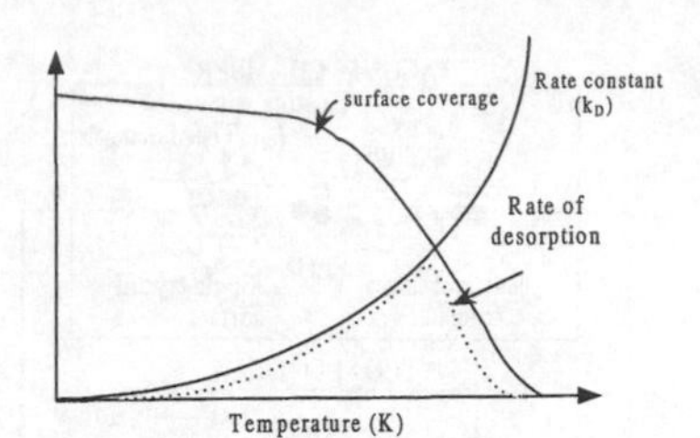

Fig. 2.53 The rate of desorption (dotted line) as a convolution of changes in surface coverage and rate constant as a function of temperature.

with temperature, the surface coverage decreases simultaneously. The observed desorption kinetics are therefore a convolution of these two factors, as illustrated in Fig. 2.53 and stated explicitly in eqn 2.39 below.

TPD spectra are usually collected as a family of curves of increasing initial surface coverage.

Subsequent analysis may yield:

(i) the activation energy for desorption (E_d);

(ii) information on the nature and strength of lateral adatom interactions; and

(iii) the relative surface coverage of adsorbate.

The rate of the desorption process may be formulated as

$$\frac{-\mathrm{d}N}{\mathrm{d}t} = k_d N^m \tag{2.39}$$

where N = the number of adsorbed molecules, k_d is the rate constant for the desorption process, and m is the order of the reaction. If one now makes the substitution

$$\frac{\mathrm{d}N}{\mathrm{d}t} = \frac{\mathrm{d}N}{\mathrm{d}T} \times \frac{\mathrm{d}T}{\mathrm{d}t} = \frac{\mathrm{d}N}{\mathrm{d}T}\beta \tag{2.40}$$

Where $\beta = \frac{\mathrm{d}T}{\mathrm{d}t}$ (the heating rate)

eqn 2.39 can be rewritten as:

$$\frac{-\mathrm{d}N}{\mathrm{d}T} = \frac{k_d}{\beta} N^m$$

and substituting for k_d from 2.38:

$$\frac{-\mathrm{d}N}{\mathrm{d}T} = N^m \frac{A}{\beta} \exp(-E_d/RT) \tag{2.41}$$

When $T = T_p$ (the thermal desorption peak maximum)

$$\frac{\mathrm{d}^2N}{\mathrm{d}T^2} = 0 \text{ (the rate of desorption reaches a maximum).} \tag{2.42}$$

Hence, by differentiating eqn 2.41 with respect to T and equating to zero, one obtains a general expression relating to T_p, E_d, and N

$$\frac{E_d}{RT_p^2} = \frac{A}{\beta} m N^{m-1} \exp(-E_d/RT_p) \tag{2.43}$$

In particular, for $m = 1$ (first-order desorption)

$$\frac{E_d}{RT_p^2} = \frac{A}{\beta} \exp\,(-E_d/RT_p) \tag{2.44}$$

Thus, as β and T_p are experimentally measurable parameters, E_d may be evaluated so long as the pre-exponential factor A is known. It is commonly assumed for $m = 1$ that A is of the same order of magnitude as the molecular vibrational frequency and is usually assumed to be 10^{13} s^{-1}. The desorption activation energy can be evaluated iteratively by initially 'guessing' E_d and refining the 'guestimate' until the equality in eqn 2.44 is satisfied. Alternatively, for $m = 1$ and A/β between 10^8 and 10^{13} K^{-1}, it has been shown that a modified version of eqn 2.44 is accurate

A useful 'rule of thumb' for estimating E_d in kJ mol^{-1} from T_p in Kelvin is to take $E_d = T_p/4$.

$$E_d = RT_p[\log_e\left(\frac{AT_p}{\beta}\right) - 3.46] \tag{2.45}$$

where $R = 8.314$ J mol^{-1}K^{-1} and T_p is in Kelvin.

As may be seen from equations 2.44 and 2.45, the desorption peak maximum is independent of adsorbate coverage (2.44 and 2.45 contain no terms in N). With increasing adsorbate coverage, the desorption peak maximum remains at a constant temperature and simply increases in intensity.

It should be stressed that the above analysis makes a number of assumptions, in particular that the activation energy and pre-exponential factor are coverage independent. In addition, it has been assumed that desorption occurs in a single step. Desorption for systems that exhibit 'precursor state' adsorption kinetics have a more complex desorption mechanism since the precursor state will also play a role in the desorption process. In such cases application of simple one-step desorption kinetics may lead to significant errors in desorption activation energies evaluated using eqn 2.45.

For $m = 2$, eqn 2.43 gives

$$\frac{E_d}{RT_p^2} = \frac{2AN}{\beta} \exp\left(-E_d/RT_p\right) \tag{2.46}$$

This means that T_p is now dependent on N for a second-order process and, in fact, as N increases, T_p is seen to decrease for a fixed value of E_d. Hence, the characteristic thermal desorption spectrum for a second-order desorption process as a function of increasing coverage is a shift in T_p to lower temperatures but the peaks remaining symmetric. This contrasts with a first-order process, whereby desorption peaks are asymmetric.

While chemisorbed monolayers generally exhibit first- or second-order desorption kinetics, multilayer systems will obey zero-order kinetics. In a multilayer system the first monolayer is bonded via a strong chemisorption bond to the substrate. In the second and subsequent monolayers, the bonding resembles that in a condensed solid of the pure adsorbate. The influence of the bonding to the surface is almost completely screened out. This weaker multilayer bonding manifests itself in a desorption peak at lower temperature (lower bond strength), relative to desorption from the chemisorbed monolayer. Figure 2.54 illustrates TPD from a W(110) surface exposed to increasing amounts of palladium. The high temperature peak is a result of desorption from the chemisorbed monolayer and exhibits first-order kinetics, while the low temperature state is associated with multilayer desorption. Multilayer desorption peaks do not saturate and will continually increase in intensity as more and more material is condensed on to the surface. The temperature for maximum desorption (T_p) for a zero-order process shifts to higher temperature with increasing coverage and all desorption curves have a common leading (low temperature) edge. The shift to higher temperature with increased thickness of the multilayer film occurs because the desorption rate increases exponentially with temperature. Thus, the rate can increase indefinitely until all multilayers have been stripped away, and the peak temperature is simply limited by the amount of multilayer material. Hence, an infinitely thick layer would exhibit an infinitely high desorption peak maximum! In summary, for adsorbate systems that do not exhibit lateral interactions, the order of

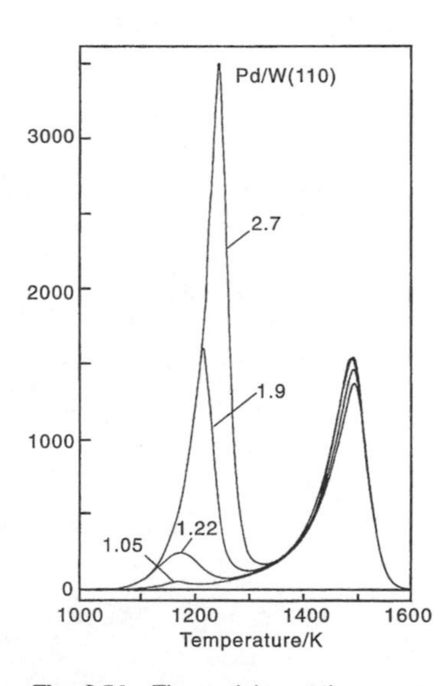

Fig. 2.54 Thermal desorption spectra for palladium on W(110) as a function of coverage up to a maximum of 2.7 monolayers. Note that the area of the high temperature peak corresponds to the desorption of the first palladium monolayer. The larger, lower temperature peak corresponds to a palladium multilayer desorption. In addition, it is evident that multilayer desorption occurs only after filling of the first monolayer. Adapted from ref. 11.

desorption can be differentiated simply by consideration of the variation in the desorption peak maximum with temperature and the peak shape.

Exceptional behaviour in TPD

An assumption of a coverage-independent activation energy for first order desorption predicts a coverage–independent desorption maximum. However, in some adsorption systems, ostensively exhibiting first order kinetics, increasing exposure may lead to the desorption peak maximum shifting to lower temperature. Furthermore, low temperature peaks can often appear at the highest exposure.

The existence of multiple desorption peaks and coverage-dependent shifts in peak maxima may arise from:

(i) the presence of more than one distinct binding site with differing activation energies for desorption (e.g. atop, hollow, etc.); and

(ii) coverage-dependent lateral interactions between adsorbates.

It is difficult to differentiate between (i) and (ii) without the aid of additional information, such as that provided by vibrational spectroscopy (RAIRS or HREELS), which has the ability to signal clearly variations in the number of binding sites (see Section 2.9).

As discussed in Chapter 1, repulsive interactions destabilize neighbouring atoms and lead to a lowering in their adsorption energy, hence producing a shift in the desorption maxima to lower temperature with increasing coverage. This is due to increasing interadsorbate repulsion as the adsorbate–adsorbate separation is reduced and the molecules are forced closer together (see Section 1.14). For example, at coverages less than 0.5 ML, CO molecules on Co(10$\bar{1}$0) occupy a single adsorption site and exhibit a single desorption peak. At coverages above 0.5 ML, CO molecules are forced to occupy nearest-neighbour sites along the close-packed rows of substrate atoms. This leads to the formation of a second low temperature desorption peak associated with the much increased repulsion between CO molecules at such short intermolecular separations.

Surface coverages from TPD

A further valuable piece of information is contained within a TPD spectrum; **the relative surface coverage**. Provided the pumping speed of the vacuum chamber remains constant during desorption experiments, the integrated area of a TPD peak is directly proportional to the surface coverage, provided that all other variables that affect the mass spectrometer signal, e.g. detector gain, distance between sample and ionisation chamber, heating rate, etc., are kept constant. Thus, simple integration of the area under two TPD curves yields the relative coverage directly. Absolute coverages may be obtained by this method provided a TPD curve from a known absolute coverage of adsorbate is available. For example, adsorption of CO on Rh (111) leads to formation of a $(\sqrt{3} \times \sqrt{3})$ R30° LEED structure in which the $\sqrt{3}$ beams maximize their intensity at exactly 0.33 ML. Thus, the integrated area under a TPD curve from this structure corresponds to a known absolute coverage and any unknown coverages may be determined by a simple ratio of integrated areas of the desorption spectra.

$$\theta = \frac{\text{area under TPD curve for unknown coverage}}{\text{area under TPD curve for known coverage}} \times \text{known coverage} \tag{2.47}$$

TPD may also be used to investigate decomposition reaction mechanisms. As adsorbed molecules may decompose into a range of products of differing mass, such an experiment requires the simultaneous monitoring of several masses ('multiplexing'). This may be achieved by allowing the mass spectrometer to rapidly (several times a second) switch between several detected masses. Thus, while truly continuous monitoring is not possible, a quasi-continuous output is obtained. For example, the decomposition of formic acid on copper(110) can be investigated via TPD obtained by monitoring the masses of the parent ion (HCOOH = 46 amu), hydrogen (H_2 = 2 amu), and carbon dioxide (CO_2 = 44 amu) as the sample temperature is ramped. Hydrogen desorption is observed below room temperature, followed by simultaneous evolution of CO_2 and H_2 at higher temperatures. As separate experiments studying CO_2 and H_2 adsorption *alone* indicate that desorption of both CO_2 and H_2 is complete below room temperature, the only explanation for the high temperature desorption peak is the decomposition of an intermediate containing carbon, hydrogen, and oxygen into the products CO_2 and H_2, which desorb immediately. This process is termed **reaction limited desorption**. A formate intermediate has been proposed. The following reaction mechanism has therefore been suggested to account for HCOOH decomposition on Cu(110):

amu = atomic mass unit.

$$HCOOH_{(ad)} \rightarrow HCOO_{(ad)} + H_{(ad)}$$
$$2H_{(ad)} \rightarrow H_{2(g)} \text{ (275 K peak)}$$
$$2HCOO_{(ad)} \rightarrow 2CO_{2(g)} + H_{2(g)} \text{ (475 K peak)}$$

If this were the case, the low temperature hydrogen desorption would be a result of deprotonation of the hydroxyl hydrogen. In order to test this hypothesis, the technique of 'isotopic labelling' may be used. Isotopic substitution, as in many other fields of chemistry, is a powerful technique to investigate reaction mechanisms. Adsorption of deuterated formic acid (HCOOD) would thus lead to deprotonation producing exclusively deuterium atoms, which upon recombination would produce low temperature D_2 evolution (mass 4) rather than H_2 (mass 2) (see Fig. 2.55). Clearly, if a HD mixture were observed in the desorption spectra a more complex mechanism involving cleavage of the C–H and O–D bonds would be required. In many cases, judicious choice of isotopically substituted molecules allows a surface reaction mechanism to be untangled and a decomposition mechanism postulated.

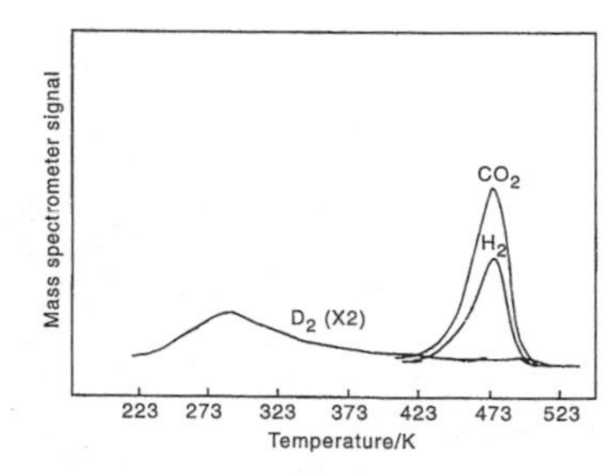

Fig. 2.55 TPD spectra of CO_2, D_2 and H_2 derived from the decomposition of HCOOD on Cu(110). Adapted from ref. 12.

2.8 Molecular beams

A 'molecular beam' is a collimated source of gas molecules of well-defined spatial distribution, particle flux, and, in certain cases, energy distribution among the internal modes of freedom (translational, vibrational, rotational, electronic). The complexity of the beam source varies considerably. The simplest 'thermal' sources give rise to a Maxwell–Boltzmann internal energy

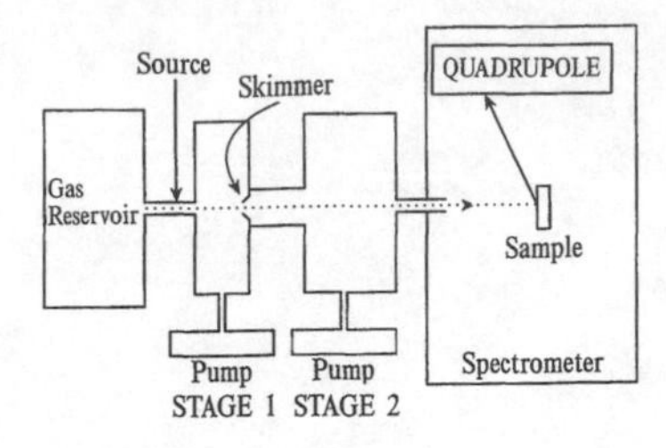

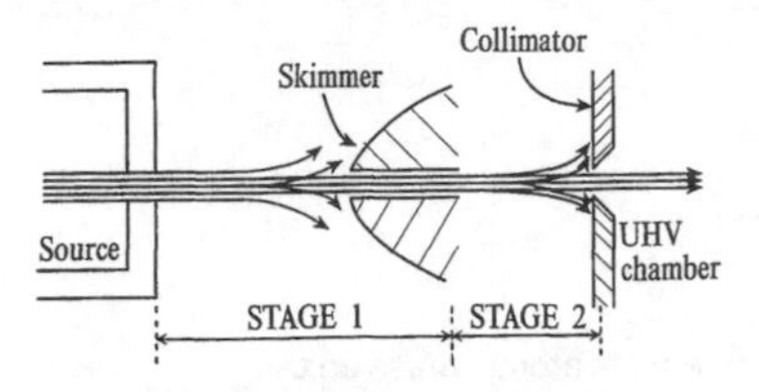

Fig. 2.56 Schematic diagram of the apparatus used in a molecular beam experiment.

distribution. The most complex involve supersonic nozzle sources combined with laser excitation in which essentially all quantum states within the gas molecule may be controlled. As mentioned in Chapter 1, the precise translational, rotational, and vibrational state of an incident molecule can profoundly affect the probability of dissociative adsorption and other surface reactions. Hence, molecular beam studies are important in the study of adsorption activation energies, sticking probabilities, and surface reaction kinetics.

Thermal beams

Figure 2.56 illustrates schematically the simplest 'thermal' molecular beam source. The beam is formed by expansion of a gas, typically at a pressure of $\approx$0.1 atmospheres in the source chamber through a Pyrex capillary of diameter $\sim$0.1 mm. The molecules collide many times with the capillary walls on their passage through, hence attaining thermal equilibrium with the capillary. A platinum heating wire wound around the source may serve to vary the incident gas temperature, allowing flexibility in controlling the average kinetic energy of the particles. As the beam must enter a vacuum chamber, differential pumping stages consisting of small intermediate chambers, each with their own vacuum pump, are inserted between the source and the main chamber. This ensures that the analysis chamber remains at ultra-high vacuum once the beam is allowed to strike the sample surface. As a molecular beam tends to increase in diameter as a function of distance from the source capillary, a 'skimmer' located in the first chamber, consisting of a small orifice, is used to recollimate the beam. Final collimation occurs in the main UHV chamber just prior to interaction with the surface under study, yielding a beam diameter of a few millimetres at the sample surface and a flux of 10^{13}–10^{14} molecules cm^{-2} s^{-1}, i.e. an arrival rate of between approximately 0.01 and 0.1 monolayers per second. Upon striking the sample surface, reflected particles are detected with a quadrupole mass spectrometer tuned to the charge to mass ratio of the beam gas or possible desorption products, and located several centimetres from the target surface.

The most basic measurement possible with a beam source is that of **adsorption rate** or **'sticking probability'** (see Section 1.6). The sticking probability (S) is an index of the ability of a surface to adsorb incident particles and is defined in eqn 1.41. S takes the value unity if every particle that collides with the surface adsorbs, and zero if no adsorption occurs. The sticking probability depends critically on the substrate–adsorbate combination, surface temperature, adsorbate coverage, and the extent of excitation of internal degrees of freedom of the incident particle (again see Section 1.6).

Figure 2.57 illustrates schematically a sticking probability measurement. The dashed curve shows how the partial pressure of gas responds when incident upon an inert surface in which all molecules are reflected (no adsorption). Note that the abrupt increase in the partial pressure at $t = 0$ from P_0 (the base pressure of gas in the UHV chamber) to P_f is immediately reversed when the beam is switched off. In addition, the area of the 'rectangle' bounded by the dashed line in Fig. 2.57 is a quantitative measurement of the total number of molecules introduced into the chamber from the molecular beam source. The full line illustrates a beam incident on a surface leading to a more gradual pressure rise in the chamber owing to some of the incident

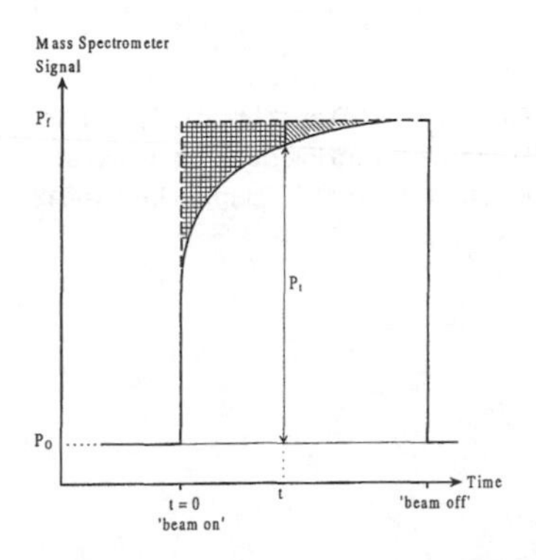

Fig. 2.57 The measurement of sticking probabilities via molecular beam data. See text for details.

particles adsorbing on to the substrate and, hence, not appearing in the gas phase. In this case, the surface is effectively acting as a fast pump. The number of molecules adsorbed is given by the total shaded area between the dashed and full curves.

The measurement shown in Fig. 2.57 essentially contains information concerning the variation of the sticking probability as a function of surface coverage. The sticking probability at any particular adsorption time (t) may be calculated using

$$S(t) = \frac{P(f) - P(t)}{P(f) - P_0} \tag{2.48}$$

where $P(t)$ is the increase in the partial pressure of impinging gas over and above P_0 at time t, and $P(f)$ is the pressure when the surface is saturated with gas. The corresponding surface coverage [$\theta(t)$] may be determined provided the beam flux (F) in molecules cm^{-2} s^{-1} is known, since

$$\theta(t) = F \int_0^t S(t)\mathrm{d}t \tag{2.49}$$

The integral in eqn 2.49 is equal to the cross-hatched area in Fig. 2.57. The data contained within Fig. 2.57 may be converted into a sticking probability *versus* surface coverage plot (see Fig. 1.7). It is stressed once again that plots of S *versus* θ can give information on the adsorption mechanism (Langmuir *versus* precursor kinetics, for example).

For many adsorbates there exists an energy barrier to sticking. This is most common in adsorption processes requiring the cleavage of one or more bonds within the molecule, (as discussed on page 13). In such cases the sticking probability is often small, because the majority of incident molecules do not have the required energy to surmount this barrier. Variation of the sticking probability with beam temperature allows the activation barrier for dissociation to be calculated. Reference to Fig. 1.8 in Chapter 1 for the dissociative adsorption of oxygen shows that as the molecule approaches the surface, it first enters a weakly bound physisorbed state. To proceed into the more thermodynamically favoured dissociated state, either *directly* or after **trapping** in the physisorbed state, it must pass over the activation energy barrier (E^{a}_{Diss}). The crossing point between the physisorption and chemisorption wells represents a state in which the oxygen molecular bond is partially broken and the surface–oxygen bonds partially formed. As transfer into the chemisorbed well is an 'activated' process whose rate per collision is given by the sticking probability for dissociative adsorption, the sticking probability must follow an Arrhenius-type response. The zero coverage sticking probability obeys the relationship

$$S_0 = S' \exp\left(\frac{-E^{a}_{Diss}}{RT}\right) \tag{2.50}$$

where T is the absolute temperature of the incident gas molecules and S' is the sticking probability in the absence of an energy barrier. Hence measurement of S_0 for a range of absolute temperatures allows the activation energy to be deduced via a plot of $\ln(S_0)$ versus $1/T$.

$$\ln(S_0) = \ln(S') - \frac{E^{a}_{Diss}}{R}\left(\frac{1}{T}\right) \tag{2.51}$$

$y = c + m(x)$

Molecular beams are also important in the study of surface chemical reactions. In general, reactions involving two or more species can be followed in two ways.

(i) Pre-adsorption of a known coverage of reactant A on a surface followed by beaming on to the surface reactant B.

(ii) Using a 'mixed beam' in which the two reactants are present in the gas beam and arrive simultaneously at the surface.

In both cases the multiplexing facility of the quadrupole detector (as discussed in the section on TPD) is crucial, allowing the simultaneous measurement of the rate of adsorption of reactants and rate of evolution of possible desorption products.

Supersonic beams

'Supersonic' beam sources are formed by expansion of a gas through an orifice in the same manner as 'thermal' beams. However, they differ fundamentally in that the source pressure is much higher (several atmospheres) and the orifice is much smaller (typically below 100 μm). This leads to the formation of a jet of gas in which the molecular velocity exceeds the local speed of sound, hence the term 'supersonic'. Because of the higher gas loads flowing from the source into the differentially pumped stages, much larger vacuum pumps are required, making supersonic beam equipment more bulky and expensive than its thermal beam counterpart. Owing to the conditions of the expansion, the velocity distribution, which initially is Maxwell–Boltzmann (very broad), becomes strongly peaked and exhibits a low temperature owing to the cooling experienced by the gas upon expansion. More importantly, all internal modes of freedom (vibrational, rotational, electronic) are frozen into their ground states, hence the exact quantum state of the molecules that form the beam are under the experimenter's control. 'Supersonic' beams combined with excitation of the beam en route to the sample using lasers tuned to a particular molecular vibrational frequency, allows the experimentalist to pump incident molecules selectively into well-defined excited rotational/vibrational/electronic states before reaction with the surface, hence allowing the energy pathways by which surface reactions proceed to be probed in a fundamental manner. In simple terms, the precise contributions to overcoming the activation energy barrier to dissociation may be resolved, using supersonic beams, into a specific molecular translation, rotation, vibration, or combination of all three. It is using such experimental techniques that a deep understanding, at a molecular level, of the mechanisms involved in surface-catalysed reactions are currently being pursued and will be explored in the coming years.

2.9 Vibrational spectroscopy

Vibrational spectroscopy is a useful tool for probing the bonding of atoms and molecules adsorbed on a surface. There are three broad regions of interest that are relevant to surface study.

Vibrations of substrate atoms within the surface layer: surface phonons

Phonons have already been mentioned in relation to the IMFP of an electron. They are quantized vibrations of the crystal lattice (all atoms are interconnected in a solid and may be thought of as acting like a giant, coupled array of simple harmonic oscillators). The surface region has its own 'phonon' spectrum associated with quantized vibrations of atoms localized specifically in the surface. Surface phonon vibrational amplitudes decay exponentially into the bulk and yield information concerning the strength of surface bonding. The phonon modes of a surface occur at very low energies, typically below 600 cm^{-1} (~75 meV). So, for example, ZnO exhibits a surface phonon at 560 cm^{-1} (70 meV), whereas phonon frequencies for metal atoms bonded at surface steps are located at even lower energies, at around 200 cm^{-1} (~25 meV).

Chemisorbed atoms

Adsorbate atoms exhibit vibrational frequencies in the range 20–100 meV (200–800 cm^{-1}).

Chemisorbed molecules

The majority of vibrations of functional groups within adsorbed molecules generally occur above 800 cm^{-1}.

Two principal vibrational techniques are used to measure the vibrational properties of surfaces. As surface chemists are particularly interested in adsorbed molecules, we shall concentrate on the vibrational spectroscopy of chemisorbed species.

3.9.1 Reflection–absorption infrared spectroscopy (RAIRS)

The fact that infrared spectroscopy provides specific information on the types of bonds present in a molecule, is non-destructive, and does not require UHV has made it a highly versatile technique for surface analysis. However, many surfaces are opaque to infrared radiation, so transmission experiments are not viable. Hence, most studies use the so-called reflection mode. Figure 2.58 illustrates a typical experimental set-up for a vacuum RAIRS experiment. Infrared radiation is focused through an IR-transparent window (usually an alkali halide) on to the sample surface at grazing incidence. The light is generally polarized prior to focusing. The sample, acting as a mirror, reflects the beam out of a second vacuum-sealed window, where it is recollimated on to a photoconductive semiconductor detector such as mercury cadmium telluride (MCT) (detection range 5000–800 cm^{-1}). Typically, the path of the IR beam external to the UHV chamber is purged with dry nitrogen to minimize interference from gas phase absorption bands associated with atmospheric H_2O and CO_2.

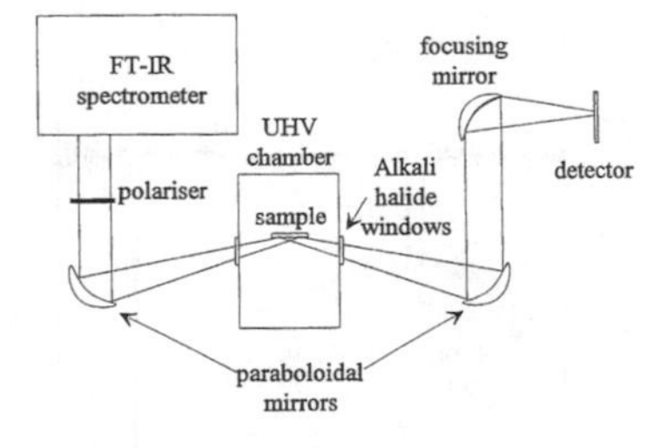

Fig. 2.58 Schematic diagram of the experimental configuration used in reflection–absorption infrared spectroscopy.

Since a monolayer of molecular adsorbate has a coverage of about 10^{15} molecules cm^{-2}, and the total area of a typical sample probed is less than a square centimetre, sensitivity is the major problem encountered in RAIRS. Hence, the experiment is performed in a grazing incidence geometry, which tends to maximize surface sensitivity for the following reasons.

First, in the reflection mode a 'double pass' geometry is used. The incident beam must pass once through the surface layer before hitting the reflecting substrate, and a second time on its outward journey to the detector. The adoption of a grazing incidence geometry also leads to a rapid increase in path length, hence increasing the sensitivity for very thin layers of adsorbate. Second, the magnitude of the electric vector of the radiation also changes drastically as the angle of incidence approaches grazing. When infrared radiation is incident on a surface, the amplitude and the phase of the radiation change upon reflection. The exact mechanism by which these changes occur is complex. However, the net result is an enhancement in the electric field vector of the IR photon ($\hat{E}$) **perpendicular** to the surface for grazing incidence geometry, and zero magnitude of $\hat{E}$ parallel to the surface. Since the intensity of the IR band will depend on $\hat{E}^2$, this means that **only molecular vibrations giving rise to a dynamic dipole moment perpendicular to the surface will yield IR absorption;** this is termed the **surface selection rule** for surface vibrational spectroscopy. Another way of rationalizing the surface selection rule is to consider the response of the valence electrons of the substrate to the molecular vibrations of an adsorbate (Fig. 2.59). It is seen that the polar CO molecule induces image charges in the substrate. In the upright configuration, the dipole moments of the image charges and the CO molecule re-enforce each other. Hence, upon vibration, a surface-enhanced change in the *net* dipole obtained gives rise to significant IR absorption. In contrast, for a molecular dipole aligned parallel to the surface, upon vibration, both the image and molecular dipoles cancel and no IR absorption is observed.

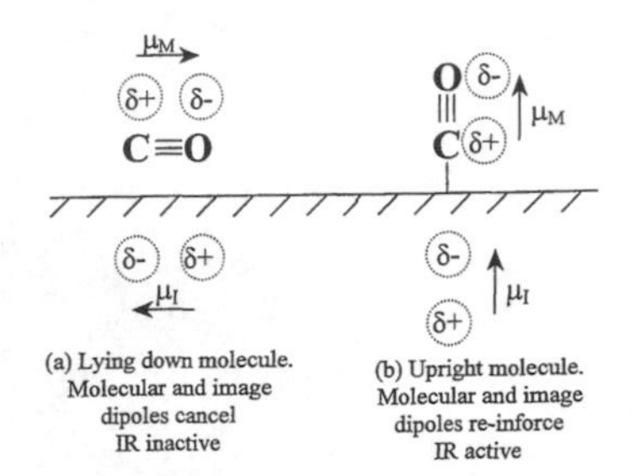

Fig. 2.59 Illustration of surface selection rule for observation of RAIRS in terms of molecular and image dipoles.

Because of the extremely high inherent resolution of RAIRS (≤ 4 cm^{-1}), inspection of the spectra for simple molecules yields important information. For example, CO molecules bonding in sites of differing coordination, e.g. atop, bridge bonded, or higher coordination sites, will give rise to different vibrational frequencies owing to the differing degree of back-donation of substrate electrons into the lowest unoccupied orbital of CO [13]. Figure 2.60 shows a series of reflection–IR spectra from Ag(111) in which ethanoic acid has been co-adsorbed with oxygen at low temperature. The spectrum at 180 K indicates the presence of two catalytic intermediates: a methoxy species and a formate species. Such assignments are usually made on the basis of direct comparisons with the stretching frequencies of known molecular species. After heating the surface to 280 K, the intense band at 1024 cm^{-1} from the methoxy carbon–oxygen stretch disappears, along with bands at higher wavenumber attributable to the OCH_3 intermediate, while the bands due to the formate remain unchanged. As no new bands are observed in the range 2100–1500 cm^{-1} (typical of carbon monoxide species), it may be concluded that the methoxy has decomposed and desorbed from the surface. Finally, heating to >350 K leads to loss of absorption bands associated with the formate species and the appearance of a strong band typical of carbon monoxide, indicating decomposition of the formate intermediate to CO. Hence, RAIRS is seen as an extremely useful method of elucidating molecular reaction mechanisms at surfaces. If ambiguities arise in the assignment of surface IR bands, particularly those associated with hydrogen, as with TPD, isotopic substitution can often be used to clarify the nature of the intermediate, since a change in vibrational frequency will result when analysing the deuterated molecular analogue.

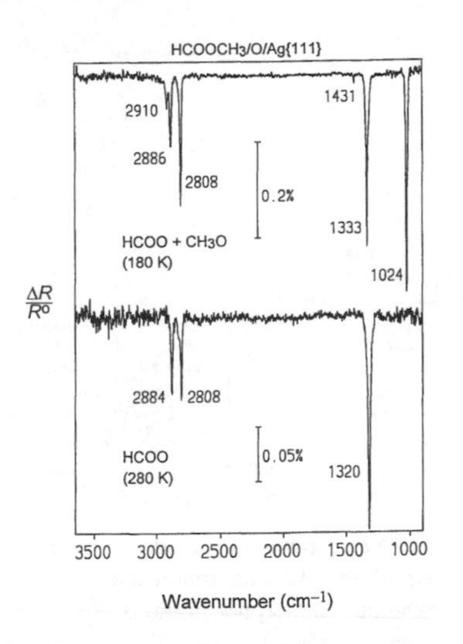

Fig. 2.60 RAIRS from ethanoic acid co-adsorbed with oxygen on Ag(111). The desorption of the methoxy intermediate is indicated by the disappearance of the peak at 1024 cm^{-1} after heating to 280 K. The formate intermediate present at 180 K remains adsorbed after heating to 280 K since negligible changes occur in the absorption bands at $\approx$1330 and 2800 cm^{-1} [14].

3.9.2 High resolution electron energy loss spectroscopy (HREELS)

A second method of exciting atomic and molecular vibrations at surfaces is with an electron beam. As the quantized excitation energies in molecular vibration range between hundredths of an electron volt and a few tenths of an electron volt, the exciting electron beam must be highly monochromatic with an energy spread < 10 meV at a typical primary beam energy of between 1 and 10 eV. The monochromatic beam suffers quantized energy losses owing to the excitation of surface vibrations, leading to energy loss peaks being observed. A typical HREELS spectrum of CO adsorbed on Rh(111) is illustrated in Fig. 2.61. Energy loss peaks appear close to the intense 'elastic' peak associated with both the metal–carbon stretch at 460 cm^{-1} and the carbon–oxygen stretches at 1855 cm^{-1} and 2065 cm^{-1}.

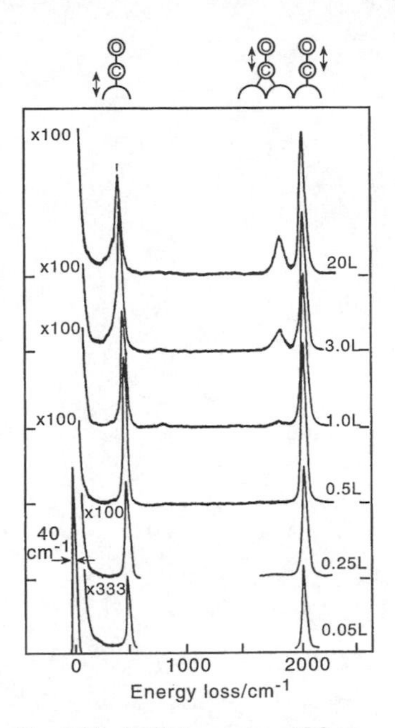

Fig. 2.61 HREEL spectra of CO on Rh(111) as a function of increasing CO coverage. Three bands are observed associated with the metal–carbon stretch (460 cm^{-1}) and two bands derived from carbon–oxygen stretches at ≈ 2000 cm^{-1} [15].

A major advantage of HREELS with respect to RAIRS is its ability to monitor vibrational modes below 1000 cm^{-1} (losses < 124 meV) associated with substrate–atom stretches. RAIRS is unable to observe low frequency vibrational modes because the IR source cannot produce a high enough flux of incident photons below 1000 cm^{-1}. In favourable circumstances, the actual adsorption site of an atom may be deduced. In the gas phase, an atom has three degrees of freedom (translation in the *x*-, *y*-, and *z*-directions) but when adsorbed on a surface these three modes are converted to **vibrations** in the *x*, *y*, and *z*-directions. These vibrational modes are sometimes referred to as 'frustrated translations'. Figure 2.62 shows the adsorption of an atom in hollow, bridge, and atop sites on either a (100) bcc or fcc surface, and the corresponding *x* and *y* frustrated translations (the *z* frustrated translation is not shown). It is evident that, for hollow and atop sites, the *x* and *y* frustrated translations are degenerate in energy and hence will occur at the same frequency. Therefore, two vibrational modes in total will be observed from atop/fourfold adsorption sites. The first would correspond to the *z* frustrated translation and the second to the two degenerate *x* and *y* frustrated translations. In contrast, for the bridge site, the *x* and *y* vibrations are clearly not degenerate, since frustrated translation in the direction of atop sites is not energetically the same as frustrated translation towards a fourfold hollow. Therefore, for bridge site adsorption, *three* vibrational frequencies should be observed corresponding to each of the *x*, *y*, and *z* frustrated translations. For the dissociative adsorption of hydrogen on W(100), three metal–hydride vibrational stretches are indeed observed in HREELS, showing unambiguously that, in this particular case, the hydrogen atom must be adsorbed in the *bridge* site.

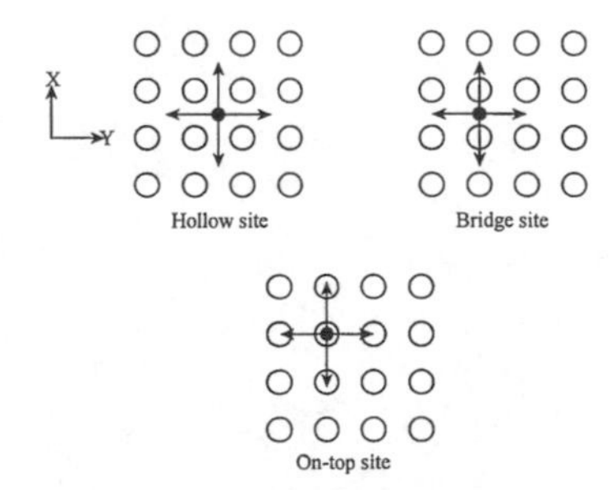

Fig. 2.62 Frustrated translations in the x- and y-directions for adsorption in atop, bridge, and fourfold hollow sites. For hollow and atop sites these vibrational modes are degenerate. This is not the case for bridge site adsorption.

A second major advantage of HREELS compared with RAIRS is that vibrational excitation may take place by two independent mechanisms.

(i) **Dipole scattering**. In this case the electric field of the incoming electron interacts with the changing electric field due to molecular vibration. The interaction is long range and hence the incoming electron senses a composite of the molecular and image dipole. Therefore, it only excites vibrations with a dipole moment change perpendicular to the surface (see Fig. 2.59). Dipole scattering is strongly peaked in the **specular** direction (angle of incidence equals angle of reflection) hence the selection rule operational in this direction is equivalent to that of RAIRS.

(ii) **Impact scattering**. This is a short-range mechanism in which a direct impact between adsorbate and electron leads to vibrational excitation. This mechanism results in a 'relaxation' of the surface selection rule such that *all* vibrations (both parallel and perpendicular to the surface) that have an accompanying dipole moment change may be excited. Vibrations excited by impact scattering are best observed **off-specular**. In this geometry, dipole-forbidden modes may be accessed without their intensity being overwhelmed by the dipole-active modes. Impact scattering is more isotropic than dipole scattering, with electrons being scattered into a wide range of different directions. This means that impact-scattered modes are, in general, considerably weaker in intensity than dipole modes, for which the entire scattered intensity is peaked in a narrow range of directions centred on specular reflection (dipole scattering).

Hence, although giving rise to somewhat broader peaks than RAIRS, HREELS has greater scope for investigating vibrational modes not normally accessible by RAIRS and, in addition, information regarding the symmetry of atomic adsorbate sites may, in favourable circumstances, be deduced.

References for Chapter 2

1. *Practical surface analysis by Auger and X-ray photoelectron spectroscopy*, eds D. Briggs and M.P. Seah, (1993), Wiley, New York.
2. K. Siegbahn, *Phil. Trans. R. Soc. Lond. A*, **318** (1986) 3.
3. K. Siegbahn in *ESCA – Atomic, molecular and solid state structure studied by means of electron spectroscopy*, (1967) Almquist and Wiksells, Uppsala.
4. K. Christmann, G. Ertl and O. Schober, *Surf. Sci.*, **40** (1973) 61.
5. Courtesy of Omicron. Vakuumphysik GmbH, Germany.
6. R.L. Gerlach and T.N. Rhodin, *Surf. Sci.*, **19** (1970) 403.
7. M. Bowker and R.J. Madix, *Surf. Sci.*, **95** (1980) 190.
8. E.W. Plummer, T. Gustafsson, W. Gudat and D.E. Eastman, *Phys. Rev*, **A15** (1977) 2339.
9. D.E. Ellis, E.J. Baerends, H. Adachi and F.W. Averill, *Surf. Sci.*, **64** (1977) 649.
10. C.J. Barnes in *The Encyclopedia of analytical science*, (1995) Academic Press, pp. 5066.
11. W. Schlenk and E. Bauer, *Surf. Sci.*, **93** (1980) 9.
12. R.J. Madix, *Surf Sci.*, **89** (1979) 540.
13. W.L. Jorgensen and L. Salem in *The organic chemists book of orbitals*, (1973), Academic Press, New York.
14. W.S. Sim, P. Gardner and D.A. King, *J. Phys. Chem.*, **100** (1996) 12509. We acknowledge Wee Sun Sim (Cambridge) and Peter Gardner (UMIST) for supplying copies of the original figures.
15. G.A. Somorjai in *Introduction to Surface Chemistry and Catalysis*, (1994) Wiley Interscience, New York.

3 Worked examples and problems

1. The pressure of O_2 gas required to give a particular coverage of adsorbed oxygen atoms on a silver surface at 700 K was 1×10^{-3} mb. However, at 800 K, a pressure of 36 mb was necessary to establish the same constant surface coverage. Estimate the isosteric enthalpy of adsorption for $O_{2(g)}$ on Ag.

Using eqn 1.34

$$[\log_e(P_1/P_2)]_\theta = \frac{\Delta H_{AD}}{R}\left(\frac{1}{T_1} - \frac{1}{T_2}\right)$$

$$\therefore \log_e \frac{36}{1} = \frac{\Delta H_{AD}}{8.314}\left(\frac{1}{800} - \frac{1}{700}\right)$$

$$\therefore 3.58 = \frac{\Delta H_{AD}}{8.314}(-1.7817 \times 10^{-4})$$

$$\therefore \Delta H_{AD} = -\frac{3.58 \times 8.314}{1.7857 \times 10^{-4}}$$

$$= -167 \text{ kJ mol}^{-1}$$

2. Test graphically the applicability of the Freundlich and Langmuir isotherms for the data tabulated alongside.
From your graph, calculate the constants of the Langmuir isotherm.

Pressure (Nm^{-2})	Mass of gas adsorbed per unit area (gm^{-2})
0.28	0.140
0.40	0.176
0.61	0.221
0.95	0.278
1.70	0.328
3.40	0.384

Freundlich isotherm—assumes logarithmic dependence of adsorption enthalpy with coverage

$$\theta = C_3 P^{1/C_4} \qquad \text{(equation 1.39)}$$

Take logs of both sides

$$\log_e \theta = \log_e C_3 + \frac{1}{C_4} \log_e P$$

But from eqn 1.19

$$\theta = m/m_\infty$$

$$\therefore \log_e \left(\frac{m}{m_\infty}\right) = \log_e C_3 + \frac{1}{C_4} \log_e P$$

$$\therefore \log_e m - \log_e m_\infty = \log_e C_3 + \frac{1}{C_4} \log_e P$$

$$\therefore \log_e m = \log_e C_3 + \log_e m_\infty + \frac{1}{C_4} \log_e P$$

$$\therefore \log_e m = \log_e (C_3 m_\infty) + \frac{1}{C_4} \log_e P$$

Form $y = c + mx$

$\therefore$ Plot $\log_e m$ *versus* $\log_e P$. If a straight line graph is obtained, adsorption obeys the Freundlich isotherm. It doesn't!

Try the **Langmuir isotherm**—assumes that enthalpy of adsorption is independent of coverage.

From eqn 1.19

Form $y = c + xm$

$$\frac{P}{m} = \frac{1}{m_\infty K} + P\left(\frac{1}{m_\infty}\right)$$

P/m (Ng^{-1})	P (Nm^{-2})
2.00	0.28
2.27	0.40
2.76	0.61
3.42	0.95
5.18	1.70
8.85	3.40

∴ Plot *P/m versus P*. If Langmuir adsorption is obeyed, the graph should be a straight line of gradient $1/m_\infty$ and intercept $1/m_\infty K$.

Since a straight line graph is obtained, Langmuir adsorption is obeyed, with $K = 1.82(\text{Nm}^{-2})^{-1}$ and $m_\infty = 0.423 \text{ g m}^{-2}$.

3. The volume of a gas (measured at 1 atmosphere pressure and 273 K) absorbed on 1 g of charcoal at various pressures was

P (cm Hg)	1.0	2.0	3.0	5.0	10.0
V (cm^3)	45.0	55.9	60.2	64.7	68.4

The diameter of the gas molecule is approximately 0.4 nm. Estimate the surface area of the charcoal. (The volume of one mole of perfect gas at 273 K and 1 atmosphere pressure is 0.0244 m^3).

Assume Langmuir adsorption! From eqn 1.20

Form $y = c + xm$

$$\therefore \left(\frac{P}{V}\right) = \frac{1}{V_\infty K} + P\left(\frac{1}{V_\infty}\right)$$

∴ Plot *P/V versus P*. If Langmuir adsorption is obeyed, the graph should be a straight line of gradient $1/V_\infty$ and intercept on the *y*-axis, $1/V_\infty K$.

From the gradient, $V_\infty = 73 \text{ cm}^3$.

Hence, since V_∞ corresponds to the volume of gas required to form a monolayer, one may convert V_∞ to the number of moles of gas

(N.B. Units—need to convert cm^3 to m^3)

$$\frac{73/10^6}{0.0244} = 2.99 \times 10^{-3} \text{ moles}$$

PV (cm Hg/cm^3)	P (cm Hg)
0.02222	1.0
0.03578	2.0
0.04983	3.0
0.07728	5.0
0.14620	10.0

Hence, the **number** of molecules corresponding to one monolayer is

$$2.99 \times 10^{-3} \times 6.023 \times 10^{23} = 1.8 \times 10^{21} \text{ molecules}$$

But, we are told that the diameter of the adsorbed gas molecule is 0.4×10^{-9} m. Hence, assuming a close-packed layer, the total area of the charcoal must be the area of **one** molecule multiplied by the total number of molecules corresponding to one monolayer:

$$\text{Charcoal surface area} = \pi \times \left(\frac{0.4 \times 10^{-9}}{2}\right)^2 \times 1.8 \times 10^{21}$$

$$= 226 \text{ m}^2$$

∴ 1 g of the charcoal has a total surface area of 226 m^2. This large figure is associated with its porosity.

4. Draw the (110) surface of a body-centred cubic material with lattice constant, 3.16 Å. Indicate the surface unit cell and calculate the surface atomic

density. How long would it take for 0.5 ML of hydrogen to adsorb on to the surface at 10^{-9} Torr at 300 K, assuming dissociative adsorption and a sticking probability of unity?

From Fig. 1.15, the unit cell is a centred rectangle of area $a \times \sqrt{2}a = \sqrt{2}a^2$.

$$\therefore \text{ area of unit cell} = \sqrt{2} \times (3.16\ \text{Å})^2 = 14.12\ \text{Å}^2$$

Each unit cell contains two atoms (the central atom is unique to the chosen cell and each of the four corner atoms contributes one-quarter of an atom, since they are shared between four unit cells. Therefore, the area per atom is $14.12\ \text{Å}^2/2 = 7.06\ \text{Å}^2$.

$$\text{Density of atoms per cm}^2 = \frac{1}{7.06 \times 10^{-16}}$$
$$= 1.416 \times 10^{15}$$

(N.B. convert Å^2 to cm^2!)

To calculate the rate of bombardment:

$$Z = \frac{p}{(2\pi mkT)^{1/2}}\ \text{cm}^{-2}\ \text{s}^{-1}$$

$$1\ \text{Torr} = 1.333 \times 10^2\ \text{Nm}^{-2}$$
$$\therefore 10^{-9}\ \text{Torr} = 1.333 \times 10^{-7}\ \text{Nm}^{-2} = 1.333 \times 10^{-11}\ \text{Ncm}^{-2}$$
$$m = \text{mass of a H}_2\text{ molecule} = \frac{2 \times 10^{-3}\ \text{kg mol}^{-1}}{6.023 \times 10^{23}\ \text{mol}^{-1}}$$
$$m = 3.321 \times 10^{-27}\ \text{kg}$$

$$Z = \frac{1.333 \times 10^{-11}\ \text{Ncm}^{-2}}{\{2\pi(3.321 \times 10^{-27}\ \text{kg})(1.381 \times 10^{-23}\ \text{JK}^{-1})(300\ \text{K})\}^{1/2}}$$
$$= 1.434 \times 10^{12}\ \text{cm}^{-2}\text{s}^{-1}$$

Therefore, a rate of coverage

$$\frac{1.434 \times 10^{12}\ \text{cm}^{-2}\text{s}^{-1}}{1.416 \times 10^{15}\ \text{cm}^{-2}} = 1.013 \times 10^{-3}\ \text{monolayers per second is expected.}$$

But, assuming **dissociative** adsorption, i.e. every H_2 molecule splitting into two atoms upon adsorption:

$$\text{Coverage} = 2.026 \times 10^{-3}\ \text{monolayers per second.}$$

Therefore, the time taken at 10^{-9} Torr to accumulate 0.5 ML of adsorbate is $0.5 \times 1/(2.026 \times 10^{-3})$ seconds or 4.1 minutes.

This example illustrates a worst-case scenario: a gas moving with a high velocity with unit sticking probability. Note that, even if the pressure is reduced to 10^{-9} Torr, a sizeable contamination has been built up in only minutes.

5. Deposition of a metal film on to a Cu substrate leads to a decrease in intensity of the Cu 60 eV kinetic energy Auger peak to 76, 50, 25, and 6% of the clean surface value for film thicknesses of 2, 5, 10, and 20 Å, respectively. Determine graphically the value of the inelastic mean free path for 60 eV Auger electrons.

An expression for the decrease in substrate Auger intensity as a function of film thickness d at normal emission is:

$$I(d) = I_0 \exp\left(\frac{-d}{\lambda}\right) \qquad \text{(equation 1.53)}$$

where I_0 = intensity of the clean substrate and λ = the inelastic mean free path. Taking natural logs

$$\ln I(d) = \ln I_0 - \frac{d}{\lambda}$$

Normalizing the clean surface intensity to 1 ($\ln I_0 = 0$)

Form $y = mx$

$$\ln I(d) = \frac{-d}{\lambda} = -\left(\frac{1}{\lambda}\right)d$$

Thus, a plot of $\ln I(d)$ *versus* the film thickness d should be linear with gradient $= -\frac{1}{\lambda}$.

d(Å)	$I(d)$	$\ln I(d)$
0	1	0.00
2	0.76	−0.274
5	0.50	−0.693
10	0.25	−1.386
20	0.06	−2.813

$$\text{Gradient} = -0.141\ \text{Å}^{-1} = -\frac{1}{\lambda}$$

$$\therefore \lambda = 7.1\ \text{Å}$$

In fact, inelastic mean free paths are determined in exactly this way, i.e. by measuring the decrease in substrate XPS or Auger peaks as a function of film thickness (which has been pre-calibrated). The analysis relies on the fact that growth occurs in a layer by layer fashion, which is often not strictly true, leading to a degree of scatter in experimental values.

6. A thin film of copper that grows layer by layer on a palladium substrate leads to a decrease in the substrate XPS peak (BE = 335 eV) to 40% of its clean surface value when excited by Al-K_α radiation ($h\nu$ = 1486.6 eV). Estimate:

(a) the Cu film thickness;

(b) the decrease in the XPS signal at an emission angle of 70°.

(You may assume that the mean atomic diameter of copper is 2.55Å. Take ϕ = 5 eV).

(a) First, calculate the inelastic mean free path of the substrate XPS peak:

$$E_{\text{kin}} = h\nu - E_b - \phi = 1486.6 - 335 - 5\ \text{eV}$$

$$E_{\text{kin}} = 1147\ \text{eV}$$

$\therefore$ The IMFP is given by equation 1.54:

$$\lambda(\text{nm}) = \frac{538a}{(E_{\text{kin}})^2} + 0.41a^{3/2}(E_{\text{kin}})^{1/2} = \frac{538}{(1147)^2} \times 0.255 + 0.41(0.255)^{3/2}(1147)^{1/2}$$

$$\therefore \lambda = 1.79\ \text{nm}$$

The decay of the substrate XPS peak with film thickness follows an exponential decay law. At normal emission, equation 1.53 holds:

$$I(d) = I_0 \exp(-d/\lambda)$$

Taking logs to the base e

$$\ln[I(d)/I_0] = -d/\lambda$$

$$d = -\lambda \ \ln[I(d)/I_0] = -1.79 \ \log_e(0.4)$$

$$\therefore d = 1.64 \text{ nm}$$

∴ The copper film thickness is 1.64 nm.

(b) For normal emission, the effective path length taken by the electron is increased by

$$d_{\text{eff}} = d/\cos\theta$$

$$\therefore d_{\text{eff}} = \frac{1.64}{\cos 70^\circ} = 4.8 \text{ nm}$$

Hence,
$$\frac{I}{I_0} = \exp\left(\frac{-4.8}{1.79}\right) = 0.068$$

∴ At an emission angle of 70°, the substrate XPS peak is 7% of its clean surface intensity. Contrast this with its value at normal emission.

7. The dissociative sticking probability of oxygen on Cu(110) was measured at two different temperatures with a thermal molecular beam. The data indicated a linear fall in S with coverage for each temperature. A LEED pattern (illustrated in the figure below) was also observed at low coverages and attained maximum intensity at a coverage of exactly 0.5ML. From this information:

(a) Calculate the activation energy for dissociative adsorption.

(b) Suggest an overlayer structure consistent with the observed LEED pattern.

(c) Explain the dependence of S with θ in terms of an adsorption mechanism.

(S_0 at 300 K = 0.22; S_0 at 850 K = 0.45 where the temperature refers to that of the thermal molecular beam.)

(a) The zero coverage sticking probability (S_0) follows an Arrhenius dependency

$$S_0 = S' \exp\left(-\frac{E_{\text{Diss}}}{RT}\right) \qquad \text{(i)}$$

where S' is the sticking probability with zero activation energy. Taking natural logs

$$\ln S_0 = \ln S' - \frac{E_{\text{Diss}}}{RT} \qquad \text{(ii)}$$

Thus, to obtain E_{Diss}, we must either have knowledge of S' or make a plot of $\ln S_0$ *versus* $\frac{1}{T}$.

If we have S_0 at two temperatures T_1 and T_2

$$(\ln S_0)_{T_1} = \ln S' - \frac{E_{\text{Diss}}}{RT_1} \qquad \text{(iii)}$$

$$(\ln S_0)_{T_2} = \ln S' - \frac{E_{\text{Diss}}}{RT_2} \qquad \text{(iv)}$$

Subtracting eqn (iii) from eqn (iv)

$$(\ln S_0)_{T_2} - (\ln S_0)_{T_1} = -\frac{E_{\text{Diss}}}{RT_2} + \frac{E_{\text{Diss}}}{RT_1} = \frac{E_{\text{Diss}}}{R}\left(\frac{1}{T_1} - \frac{1}{T_2}\right)$$

$$\ln\frac{(S_0)_{T_2}}{(S_0)_{T_1}} = \frac{E_{\text{Diss}}}{R}\left(\frac{1}{T_1} - \frac{1}{T_2}\right)$$

But $S_0 = 0.22$ at 300 K and $S_0 = 0.45$ at 850 K.

$$\therefore \ln\left(\frac{0.45}{0.22}\right) = \frac{E_{\text{Diss}}}{R}\left(\frac{1}{300} - \frac{1}{850}\right)$$

$$\therefore E_{\text{Diss}} = 2.77 \text{ kJ mol}^{-1}$$

● =Substrate spots
○ =Overlayer spots

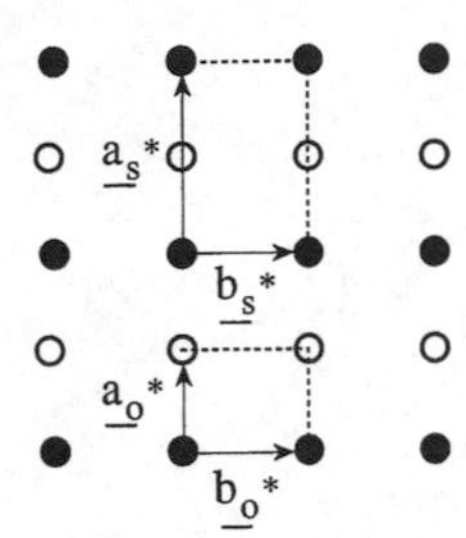

Reciprocal space

(b) From the LEED pattern (see figure opposite), determine the substrate reciprocal space unit cell and the overlayer unit cell.

Substrate unit cell vectors

$$\boldsymbol{a}_0^* = \frac{1}{2}\boldsymbol{a}_s^* + 0\boldsymbol{b}_s^*$$

$$\boldsymbol{b}_0^* = 0\boldsymbol{a}_s^* + \boldsymbol{b}_s^*$$

Form the reciprocal space matrix G^*

$$G^* = \begin{bmatrix} 1/2 & 0 \\ 0 & 1 \end{bmatrix}$$

Converting from reciprocal to real space

$$G = \frac{1}{\det G^*}\begin{bmatrix} G_{22}^* & -G_{21}^* \\ -G_{12}^* & G_{11}^* \end{bmatrix} \quad \text{(see equation 2.25)}$$

$$G = \frac{1}{1 \times 1/2 - 0 \times 0}\begin{bmatrix} 1 & 0 \\ 0 & 1/2 \end{bmatrix} = \begin{bmatrix} 2 & 0 \\ 0 & 1 \end{bmatrix}$$

Write the real space vectors of the overlayer in terms of those of the substrate

$$\boldsymbol{a}_0 = 2\boldsymbol{a}_s + 0\boldsymbol{b}_s$$
$$\boldsymbol{b}_0 = 0\boldsymbol{a}_s + 1\boldsymbol{b}_s$$

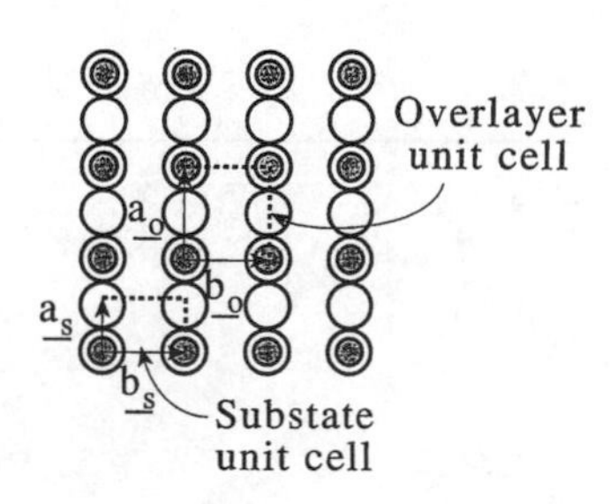

Real space

Draw the substrate surface and construct the overlayer unit cell using the above vectors (see figure opposite).

Wood's notation: Cu(110) – p (2×1) – O – (0.5 ML)

Matrix notation: Cu(110) $-\begin{bmatrix} 2 & 0 \\ 0 & 1 \end{bmatrix}$ – O – (0.5 ML)

(c) Generally, dissociative adsorption requires two adjacent sites for adsorption and hence displays a sticking probability dependence given by

$$S \propto (1 - \theta)^2$$

which is obviously non-linear in coverage. (See Figure 1.7).

However, while adsorption is dissociative, a linear dependence is observed. This is due to the adsorption mechanism and may be explained if overlayer islands of p(2×1) symmetry form even at low coverage, as shown in the figure. This explains why a p(2 × 1) LEED pattern is observed even at coverages less than 0.5ML. As the area free of p(2 × 1) islands scales linearly with coverage, between zero and 0.5ML, the linear variation in sticking probability observed experimentally is also explained. Thus, expect $S \propto (1 - 2\theta)$.

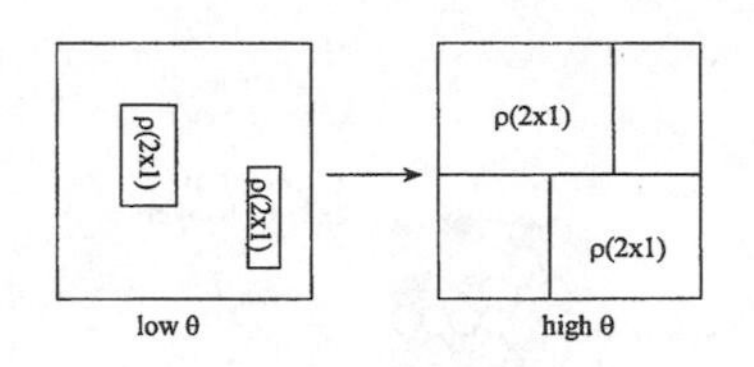

8. Adsorption of 0.25 ML of sulfur on Ni(100) leads to a work function increase of 0.24 eV. Estimate the direction and the degree of charge transfer between sulfur and nickel (nickel is face-centred cubic with lattice constant 3.51 Å). In addition, the sulfur–nickel interlayer spacing was determined by LEED to be 1.3 Å). Assume that this represents the distance between the charge on the adsorbate and the image plane.

The work function change $\Delta\phi$ is related to the dipole moment (μ) per adsorbate by the equation

$$\frac{\Delta\phi}{e} = n\frac{\mu}{\varepsilon_0}$$

|a| = 3.51Å/√2
|b|
|a| = |b| = 2.482 Å

where $n =$ the adsorbate surface density in adatoms m^{-2} and $\varepsilon_0 =$ permittivity of free space $= 8.854 \times 10^{-12}$ CV^{-1} m^{-1}. Calculating n (see figure)

Area of unit cell = $(2.48 \text{ Å})^2 = 6.15 \text{ Å}^2$

The unit cell contains one surface atom since each of the four atoms comprising the unit cell is shared between four other unit cells.

δ-
Adatom
d′
Surface
δ+
Image charge

$$\therefore \text{ No. of atoms per m}^2 = \frac{1 \text{ m}^2}{6.15 \text{ Å}^2} = \frac{1 \text{ m}^2}{6.15 \times 10^{-20} \text{ m}^2}$$

$\therefore$ One monolayer $= 1.626 \times 10^{19}$ **atoms m^{-2}**

But the surface coverage of sulfur is only 0.25 ML

$$\therefore n = \frac{1.626}{4} \times 10^{19} = 4.065 \times 10^{18} \text{ m}^{-2}$$

Calculating μ

$$\mu = \frac{\Delta\phi\varepsilon_0}{en} = \frac{(0.24 \text{ V})(8.854 \times 10^{-12} \text{ CV}^{-1} \text{ m}^{-2})}{4.065 \times 10^{18} \text{ m}^{-2}}$$

$$\mu = 5.227 \times 10^{-31} \text{ C m}$$

Now $\mu = qd$

where $q =$ charge on the adatom (see figure)

$$d = 2d' = 1.3 \text{ Å} = \text{interlayer spacing}$$

$$q = \frac{\mu}{d} = \frac{5.227 \times 10^{-31} \text{ C m}}{1.30 \times 10^{-10} \text{ m}}$$

$$q = 4.02 \times 10^{-21} \text{ C}$$

The charge on an electron, e, is equal to 1.6×10^{-19}C.

Hence, the fractional charge on the sulfur adatom is

$$\frac{4.02 \times 10^{-21}}{1.60 \times 10^{-19}} = 0.025$$

i.e. 2.5% of an electronic charge. Because the value $\Delta\phi$ has *increased*, this excess charge is negative and resides on the sulfur.

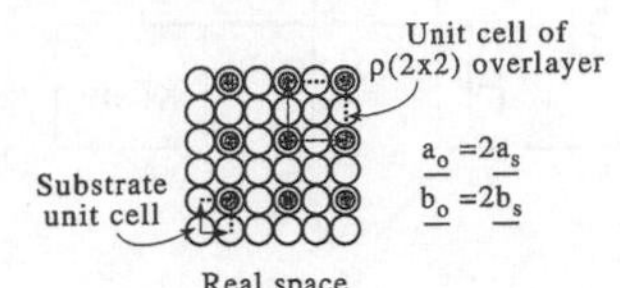

9. Calculate the number of LEED beams back-scattered from a p(2 × 2) overlayer on Pd(100) at 50 eV under conditions of normal incidence (palladium is face-centred cubic with lattice constant 3.89 Å). What angle does the (1, 0) beam make with the surface normal?

First we draw the real space structure (see Figure)

Calculating $\boldsymbol{a}_s$ and $\boldsymbol{b}_s$

$$2(|\boldsymbol{a}_s|)^2 = c^2$$

where c = lattice constant of fcc metal (see Fig. 1.14)
As c = 3.89 Å for Pd

$$|\boldsymbol{a}_s| = \sqrt{\frac{(3.89)^2}{2}} = 2.751\ \text{Å} = |\boldsymbol{b}_s|$$

Hence, $|\boldsymbol{a}_0| = 2 \times 2.75\ \text{Å} = 5.501\ \text{Å} = |\boldsymbol{b}_0|$.
The reciprocal lattice of the p(2 × 2) overlayer must now be constructed.

$$|\boldsymbol{a}_0^*| = \frac{2\pi}{|\boldsymbol{a}_0|} = \frac{2\pi}{5.501\ \text{Å}} = 1.142\ \text{Å}^{-1}$$

$$|\boldsymbol{b}_0^*| = \frac{2\pi}{|\boldsymbol{b}_0|} = \frac{2\pi}{5.501\ \text{Å}} = 1.142\ \text{Å}^{-1}$$

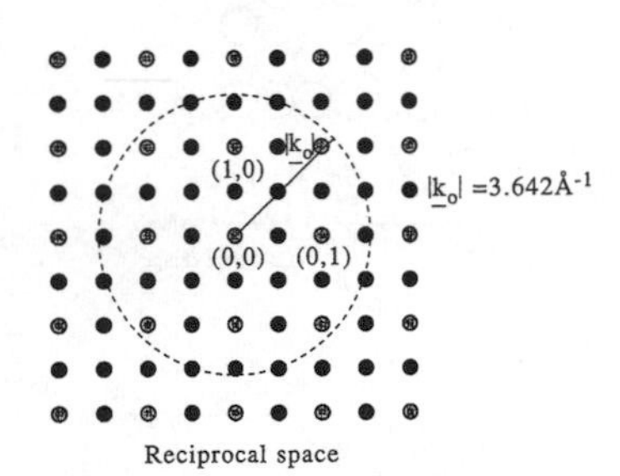

● Extra spots associated with adsorbate
◉ Clean substrate

$|\mathbf{a}_s^*| = |\mathbf{b}_s^*| = 2\pi/2.751\ \text{Å} = 2.284\ \text{Å}^{-1} = 2|\mathbf{a}_0^*|$

The reciprocal lattice is now drawn to scale (see figure opposite)

The reciprocal space lattice consists of lattice rods perpendicular to the plane of the paper. To calculate the number of diffracted beams, first compute the wavevector of the incident beam

$$|\boldsymbol{k}_0| = \frac{2\pi}{\lambda}$$

To calculate the de Broglie wavelength (λ):

$$\lambda(\text{Å}) = \sqrt{\frac{150.4}{E_p(\text{eV})}} = \sqrt{\frac{150.4}{50}}$$

$$\lambda = 1.734\ \text{Å}$$

$$\text{Hence, } |\boldsymbol{k}_0| = \frac{2\pi}{\lambda} = 3.624\ \text{Å}^{-1}$$

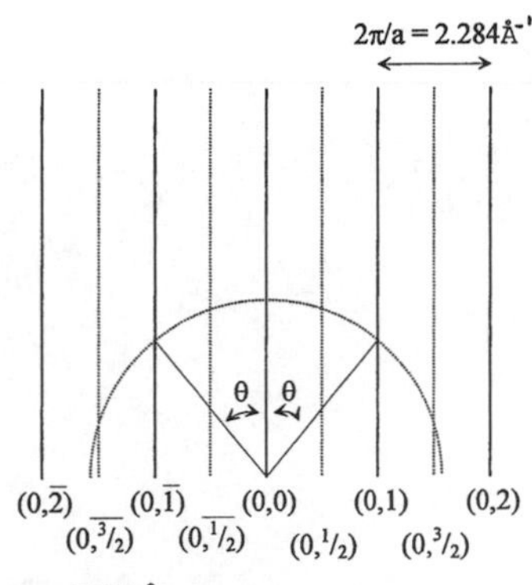

a = 2.751 Å

Draw a vector of magnitude 3.624 Å^{-1} to the origin of reciprocal space and construct a circle of radius $|\boldsymbol{k}_o|$. All reciprocal lattice points falling inside the circle will give rise to diffracted beams.

∴ We would expect to generate 37 diffracted beams at 50 eV.

To calculate the direction of the (0, 1) beam, draw a cut through the reciprocal lattice along the $[01\bar{1}]$ direction and perpendicular to the crystal surface (see figure)

$$\sin\theta = \left(\frac{\frac{2\pi}{a}}{3.624\ \text{Å}^{-1}}\right) = 0.630$$

$$\therefore\ \theta = 39°$$

In addition, it is evident that seven beams will emerge into the vacuum in this particular direction ($\boldsymbol{k}_o$ cuts the reciprocal space rods at seven positions).

10. Adsorption of 0.5ML of K on Co{10$\bar{1}$0} leads to the formation of the LEED pattern shown alongside. Saturation of the K-covered surface with CO at 300 K leads to no change in the symmetry of the LEED pattern, although strong changes in the $I(V)$ spectra of overlayer beams is observed. RAIRS indicated a single intense CO stretching vibration at 1732 cm^{-1} (for CO on clean Co(10$\bar{1}$0) at 0.5 ML, a single band is seen at 2020 cm^{-1}). The mass 28 spectrum for 0.5 ML of CO on clean Co{10$\bar{1}$0} yielded a single desorption peak at 398 K, while co-adsorption with K gave rise to an upward shift of the desorption maximum to 611 K but no change in the integrated area under the TPD peak (β = 2 K s^{-1}). Using the data outlined above:

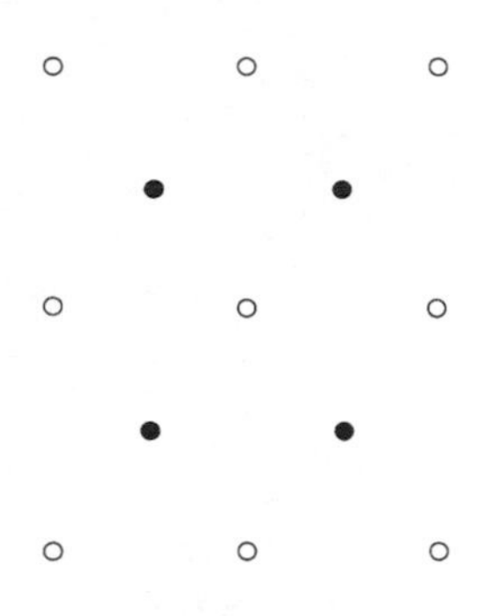

(a) suggest an overlayer structure for the CO/K co-adsorbate system;

(b) estimate the K-induced increase in binding energy of CO; and

(c) given that potassium is an electron donating adsorbate, suggest a reason for the downward shift in the CO stretching vibration induced by potassium.

From the LEED pattern, the unit cell of the K/CO co-adsorbate structure can be obtained (see figure). Writing the reciprocal space unit cell vectors of the overlayer in terms of the substrate

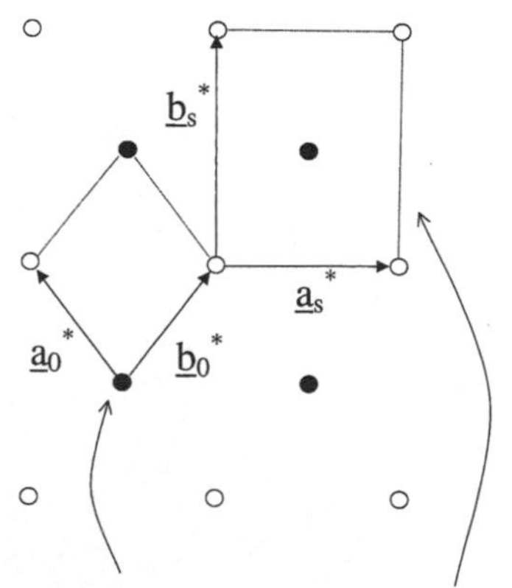

$$\boldsymbol{a}_0^* = -\frac{1}{2}\boldsymbol{a}_s^* + \frac{1}{2}\boldsymbol{b}_s^*$$
$$\boldsymbol{b}_0^* = \frac{1}{2}\boldsymbol{a}_s^* + \frac{1}{2}\boldsymbol{b}_s^*$$
$$\therefore G^* = \begin{bmatrix} -\frac{1}{2} & \frac{1}{2} \\ \frac{1}{2} & \frac{1}{2} \end{bmatrix}$$

$\therefore$ The matrix of the real space structure is given by:

$$G = \frac{1}{\det G^*}\begin{bmatrix} \frac{1}{2} & -\frac{1}{2} \\ -\frac{1}{2} & -\frac{1}{2} \end{bmatrix} \quad \text{(see equation 2.25)}$$

where $\det G^* = -\frac{1}{2}\cdot\frac{1}{2} - (\frac{1}{2}\cdot\frac{1}{2}) = -\frac{1}{2}$

$$\therefore G = \begin{bmatrix} -1 & 1 \\ 1 & 1 \end{bmatrix}$$
$$\therefore \boldsymbol{a}_0 = -\boldsymbol{a}_s + \boldsymbol{b}_s$$
$$\boldsymbol{b}_0 = \boldsymbol{a}_s + \boldsymbol{b}_s$$

Draw the substrate surface choosing a hollow site for each K atom (see figure).

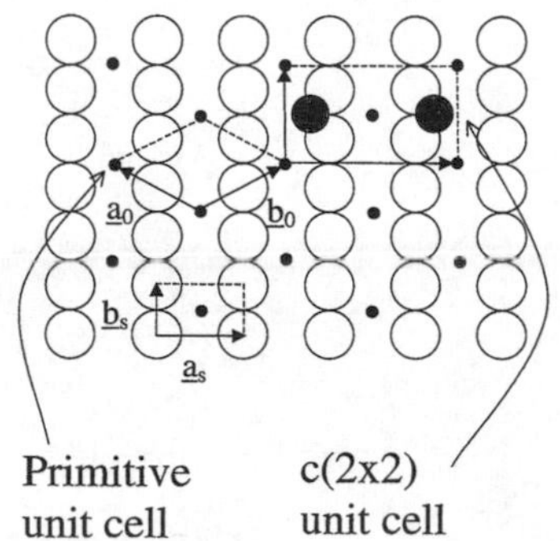

The structure as drawn gives a potassium coverage of 0.5ML. The $c(2 \times 2)$ unit cell contains two potassium atoms (the atom in the centre of the unit cell plus the atoms at the four corners of the cell). The cell also contains four substrate atoms. Therefore, the potassium coverage is 2:4 or 0.5 ML.

As the integrated area under a desorption trace is directly proportional to the surface coverage, we may say that the CO coverage in the co-adsorbate structure is also 0.5 ML.

Furthermore, the fact that only a single CO stretching vibration is observed suggests that all CO molecules are in identical chemical environments.

In addition, both the large red shift in the CO stretching frequency and the large shift in the TPD maximum are suggestive of a strong potassium–CO interaction, i.e. the K and CO are in close contact. Hence, we must place two CO molecules in the $c(2 \times 2)$ unit cell in identical chemical environments. The CO adsorption sites that offer the most 'space' (least repulsive interactions) are the 'off-bridge' sites (see figure). The only means of finding the actual CO and K sites (remember, we have *assumed* that K resides in a fourfold hollow) is a full dynamical LEED calculation.

The final conclusion we can make about the structure concerns the orientation of the CO. As a strong CO adsorption band is observed, use of the surface selection rule allows us to exclude an adsorption geometry in which the molecule is 'lying down' with the intermolecular bond axis parallel to the surface.

Assuming the CO desorption is first order, we can apply eqn 2.45 to estimate E_d:

$$E_d = RT_p[\log_e\left(\frac{AT_p}{\beta}\right) - 3.46]$$

where $R = 8.314$ J mol^{-1} K^{-1}; T_p = temperature of desorption rate maximum; β = heating rate in K s^{-1} = 2K s^{-1}; and A = pre-exponential = 10^{13}s^{-1}.

CO on clean Co{10$\bar{1}$0}

$$E_d = (8.314 \text{ J mol}^{-1} \text{ K}^{-1})(398 \text{ K})\left[\ln\left(\frac{10^{13} \text{ s}^{-1}\ 398 \text{ K}}{2\text{K s}^{-1}}\right) - 3.46\right]$$

$$E_d = 3.31 \text{ kJ mol}^{-1}[\ln(1.99 \times 10^{15}) - 3.46]$$

$$E_d = 105.1 \text{ kJ mol}^{-1}$$

CO on K-doped Co{10$\bar{1}$0}

$$E_d = (8.314 \text{ J mol}^{-1} \text{ K}^{-1})(611 \text{ K})\left[\ln\left(\frac{10^{13} \text{ s}^{-1}\ 611 \text{ K}}{2\text{K s}^{-1}}\right) - 3.46\right]$$

$$E_d = 5.08 \text{ kJ mol}^{-1}[\ln(2.04 \times 10^{15}) - 3.46]$$

$$E_d = 161.5 \text{ kJ mol}^{-1}$$

This large increase in E_d induced by potassium may either be a result of increased binding of CO to cobalt, or due to an attractive lateral interaction between K and CO (or a combination of both!).

Since back-donation of charge into the $2\pi^*$ orbitals of adsorbed CO will weaken the C–O bond, a model consistent with the data would be K donating charge into the substrate (hence itself becoming positively charged) and the substrate back-donating some of this 'extra' charge into the $2\pi^*$ orbitals on CO, leading to both a decrease in the CO stretching frequency and an enhanced electrostatic attraction between the excess negative charge on the CO and excess positive charge on the K (increase in E_d).